D0854205

The Englishwoman's Wardrobe

For
Sally Crewe
alias
Sarah Spencer

FIRST PUBLISHED IN 1986 BY CENTURY HUTCHINSON LTD
BROOKMOUNT HOUSE, 62–65 CHANDOS PLACE, COVENT GARDEN,
LONDON WC2N 4NW

CENTURY HUTCHINSON PUBLISHING GROUP (AUSTRALIA) PTY LTD
16–22 CHURCH STREET, HAWTHORN, MELBOURNE, VICTORIA 3122

CENTURY HUTCHINSON GROUP (NZ) LTD
32–34 VIEW ROAD, PO BOX 40-086, GLENFIELD, AUCKLAND 10

CENTURY HUTCHINSON GROUP (SA) PTY LTD
PO BOX 337, BERGVLEI 2012, SOUTH AFRICA

BOOK DESIGN BY BOB HOOK

SET IN GARAMOND ORIGINAL AND BAUER BODONI BLACK
PRINTED AND BOUND IN THE NETHERLANDS
BY ROYAL SMEETS OFFSET BV WEERT

BRITISH LIBRARY CATALOGUING IN PUBLICATION DATA

Huth, Angela
The Englishwoman's wardrobe: Twenty-five
Englishwomen talk about their clothes.
1. Costume – England – History – 20th century
1. Title
391'.2'0942 GT738

ISBN 0-7126-1297-1

The Englishwoman's Wardrobe

Angela Huth

Photographs by Kenneth Griffiths

CENTURY

LONDON MELBOURNE AUCKLAND JOHANNESBURG

CONTENTS

Acknowledgements

In writing *The Englishwoman's Wardrobe*, my first work of non-fiction, I came to appreciate how necessary it is to have the support of a disparate team of people whose encouragement, quite apart from their practical help, makes the peculiar hazards of non-fiction much easier for the author. I would like particularly to thank the following people: Gail Rebuck and Isabella Forbes at Century Hutchinson, who were constant in their enthusiasm, encouragement and efficiency. Kenneth Griffiths, whose beautiful photographs are in demand all over the world, was noble in his persistence in squeezing English wardrobes in between exotic foreign assignments, with the help of his assistant, Julian Broad. Ken's secretary, Maureen Bowie, displayed nothing but calm and patience while taxed by matters concerning appointments with the wardrobes. Much gratitude, of course, goes to all the contributors who gave me considerable time and thought in their uniformly busy lives, and I apologize for all the trouble I put them to in trying to think of Englishwomen whose clothes they admire. On the home front I would not have been able to cope without the invaluable assistance of my sister, Patricia, and Mrs Hey, who worked many hours overtime so that I could travel about or keep at my typewriter. Finally, I would like to thank James, Candida and Eugenie for putting up with a somewhat frazzled wife and mother while working on this book.

Introduction

Readers of this book, whose reason for buying it may be a slight but humorous curiosity about why Englishwomen dress as they do, will be familiar with the Epitome of the English Wedding Guest who inspired me to write it. She was a middle-aged lady whose smile indicated how pleased she was to have the opportunity to dress up. Perversely, making the least of that opportunity, she had plucked her 'Smart Uniform for Special Occasions' from the cupboard and, possibly, added a new hat – for, like millions of British women, hats to her were only synonymous with funerals and weddings. The uniform scarcely needs describing: patent court shoes, flesh-coloured tights, black pleated crêpe skirt, plum velvet jacket from whose neck sprouted the inevitable frills of a silk shirt, rising to meet the frills of rather too long greying hair. A good diamond rose trembled on the velvet lapel, echoed by trembling ostrich feathers on the brink of her wide-brimmed hat. (Was this matching trembling the result of ingenious forethought, or merely happy accident?) It was all pretty expensive, but hardly elegant – a muddle in front, signs of strain across the shoulders and a wisp of hair slinking down the collar at the back. (Again, like most Englishwomen, she had given no thought to her back view. Members of the Royal Family, ever alert to unkind photographers, are perhaps the only exceptions to this rule.) Why, I wondered, had this woman chosen to dress in these particular clothes? What private philosophy lay behind her decision? Was she pleased with the result? – I did not know her, and my questions were never answered. But at that moment I determined to put them to others, for surely within every wardrobe there lie fascinating explanations.

I have to admit that for as many years as I can remember I have been an automatic observer of other women's clothes. This is not because I have any particular interest in fashion: like the majority of people in this book, I am aware of which way designers are going, but take little notice if I don't like it or it doesn't suit me. But I am a natural scrutinizer, and possessor of a peculiar sort of memory that recalls, many years later, exactly what someone wore on a particular occasion. This daily observation, in England, causes few opportunities to marvel or to admire: as we all know, the elegant woman, or indeed the woman who is simply pleasing to the eye, is a very rare creature in our native land. Indeed it is hard not constantly to wonder: 'How on earth *could* she have chosen that . . .?' *The Englishwoman's Wardrobe*, as anyone will see at a glance, is as little to do with fashion as is the average Englishwoman. It provides, rather, answers to a few questions about what lies behind the way some ordinary, active women dress. How much time and thought do they give to their clothes? Why do they choose the shapes and colours they do? Who do they dress for? What is their attitude to current fashion?

I have chosen a disparate group of women to answer such questions, the busyness of their lives being almost the only thing they have in common. But even in so fractional a cross-section it is interesting to see on how many things they agree – some indication, perhaps, as to the general feeling among Englishwomen towards sartorial matters. For instance, with the exception of Maria Aitken who has to dress up for work, they all love dressing up for a special occasion. A majority are anti synthetic materials, and have great trouble in finding satisfactory shoes. Despite all the advice in women's magazines about planning a wardrobe, most of them are happily disorganized in the way they go about shopping for their clothes. Impressively loyal to old clothes, they recycle and preserve them for years. There is little interest in accessories, and hat-lovers are in a minority. As for opinions about the Englishwoman's clothes in general, the unanimous view is rotten. 'Not adventurous', 'dowdy', 'terrible', 'so often displeasing to the eye', 'frightfully drab', 'lacking flair' and 'abysmal' are some of the many disparaging observations. Elizabeth Sieff's 'better than they used to be' was the single note of faint praise. However, with one exception, the general opinion was that Englishwomen *can* look good – at grand balls and sporting occasions. By sporting occasions I think my contributors had in mind racing at Ascot and Cheltenham, more than football matches and sheepdog trials, where the horrible anorak is so popular. As for grand balls –

well, perhaps ball dresses, along with tweeds, are the British *forte*, and certainly the English 'Scotch-mist shoulders' as Jilly Cooper describes them, rising from a foam of paper taffeta ball dresses, are a magnificent sight in a stately home. Less grand evening occasions, though, are less impressive: according to Jenny Anderson, 'at Ladies' Nights and tennis dances the apparel you see defies descriptions'. Indeed, any *group* of Englishwomen is hardly cheering to the eye. Summer garden parties, for instance, bring forth all that is worst about matching trims and unwise frills. While left-wing women's clothes indicate that to appear attractive is despicable, those of the right wing often carry neatness and tidiness to the point of searing dullness. There are few sights more depressing than a gathering of academics' wives (except, perhaps, a clump of councillors' wives). In Oxford, to put on a simple silk dress for dinner is to be made to feel foolishly overdressed. As Alexander Chancellor recently observed in the *Sunday Telegraph*: 'In this country at least there is a tendency to identify scruffiness with commitment and integrity', which is sadly true, and an unenlightened attitude. As many people in this book have pointed out, it is a courtesy to others to try to look your best.

My disparate group were as united in the reasons they gave for the Englishwoman's lack of talent at dressing as they were in their condemnations: definitely nothing to do with *money*, but a lack of taste, interest, skill at putting things together and, above all, a lack of time. Also, they say, our priorities are different from those of chic Continental women: planting the herbaceous border, stoking the Aga or tending to the children are of more

Right: It took me two years to find this striped paper taffeta; eventually discovered at Allans of Duke Street. Made by a dressmaker, the dress is typical of my own romantic, vaguely Edwardian designs: with small waists and peplums. As I can never find evening shoes I like, I have them made by the theatrical cobbler Savva – variations on 18th c shapes and decorated with paste buckles from antique shops.

concern to us than what we are wearing. There is much sympathy for this view, though a touch of impatience at the thought that *both* cannot be achieved. It was generally felt that the young dress with more originality and more interestingly than the middle-aged, but perhaps they are less weighed down by that wardrobe-dampner of responsibility. While all the women in this book found considerable difficulty in thinking of someone other than the Princess of Wales whose clothes they admire, none of them hesitated when it came to describing what sort of dresser they considered themselves to be: classical, romantic, nostalgic, comfortable, show-stopping, flamboyant and sexy are some of their adjectives. Not one of them said elegant, not one of them said dull.

Clothes, I discovered, are not a subject of conversation among busy women. Each of my contributors' schemes for replenishing their wardrobes was a private matter worked out, with varying degrees of thought, alone. (Most of them admitted to dressing to please themselves, though some concessions were made to the tastes of husbands and lovers.) I felt almost prurient asking them to show me the contents of their wardrobes, for exposing one's clothes to a friend or stranger is not a normal habit, and wardrobes, after all, contain secret hopes, fears and mistakes among the pleasures and successes. As doors were opened I was regaled with explanations, justifications, apologies and only occasional snippets of modest pride or satisfaction (usually about the great age of some garment). Now, having seen hundreds of private clothes, although I can't help feeling British wardrobes indicate British restraint – a lack of exuberance, some fear of the 'touch of drama' that Jane Stevens declares is needed to be well dressed – there are also plenty of signs that the Englishwoman is not *wholly* dowdy. At least her clothes display a spirited sense of individuality – she will not be dictated to, bossed about, persuaded to wear the latest thing should it not appeal. Also, she can never be accused of being too serious about clothes. With a wise sense of priority, she puts her wardrobe into the second division of important things in her life – surely its rightful place. And when called upon to justify her attitude, to explain her clothes, she is as full of common sense as she is of humour. No wonder she dresses to please herself, and if her clothes provide entertainment for others, then that is her pleasure as well as theirs. Behind Englishwomen's clothes hang more fascinating theories than I had ever imagined, and I can only be grateful to those in this book who were so honest and revealing on the subject of their wardrobes.

Her Royal Highness the Princess Margaret

'I always, always have to be practical. I can't have skirts too tight because of getting in and out of cars and going up steps. Sleeves can't be too tight either: they must be all right for waving.'

Suit in crushed velvet with a silk shirt by Sally Crewe of Sarah Spencer, who makes to order for the Princess. This is part of the Princess's collection of 'working clothes', most of which consists of suits, or day dresses with long coats.

Members of the Royal Family, subjected to the constant, critical flare of publicity, can never get by, like the rest of us, with throwing on an inconsidered garment and hoping it won't be noticed. It will. For the Royal Family are always on show, always vulnerable to criticism, always having to think about what clothes will work. Princess Margaret, the first to admit that clothes are important to her but not one of her greatest priorities, sets about her wardrobe with practical zest, unfailing stamina and considerable humour.

'The careful planning, the fittings – they're all part of one's life,' says the Princess. 'You even get used to all those fittings – after a while, it's a bit like going to the dentist. I've learned never to have too long a fitting or you begin to droop. I plan twice a year: summer things and winter things. Once that's over I put it all out of my mind, unless I suddenly realize I haven't the right thing for the right occasion. Then I just have to try to fudge it, try to get away with putting things together that I already have.

'I can rarely dress for fun,' goes on the Princess. 'I have very few home clothes, mostly working clothes. My working clothes are like most people's best clothes. I wear last year's for some private occasions, but they're too grand for the country. I always, always have to be practical. I can't have skirts too tight because of getting in and out of cars and going up steps. Sleeves can't be too tight either: they must be all right for waving. Then I can't have anything that crushes too much – linen, for example. Back views are very important. Most people don't think about their backs, but obviously, as I'm going to be seen from all directions, I have to. I have to have things that photograph well: there's no use in having a really pretty dress which doesn't photograph. I have to think about change of temperature: long coats and dresses are more useful than suits. You can take the coat off when you arrive inside and be dressed up underneath. But you can't very well take off the coat of a suit and go about in just a shirt

Carl Toms, the renowned theatrical designer and great friend of the Princess, occasionally designs something especially for her. This was for the ball the Queen gave at Windsor in 1985 to celebrate the clutch of twenty first birthdays in the Royal Family.

on an official occasion. So you see I'm limited by such things, pretty restricted.'

Despite considerable additions to her wardrobe every year, the Princess likes to keep old favourites for a long time, and wears them often. She has a facecloth cloak, for instance, of brilliant peacock blue, made for her at Sarah Spencer. '*So* useful. I wear it over and over again. I hope it will go on and on.' Looking back, she vividly remembers many favourite dresses (several of which she has donated to the Victoria and Albert Museum). They include a pink and grey dress made by Jean Dessés in the fifties: the jacket could be taken off to transform it into a short evening dress with straps. With ruched tulle across the bodice and a single layer of black net over the whole thing, it shimmered with the kind of complicated subtlety characteristic of haute couture clothes of that time. Another much loved dress, in which the Princess was photographed on her eighteenth birthday, was of pale blue grosgrain, made by Norman Hartnell: the epitome of the New Look, which Princess Margaret particularly loved. Hartnell ('always so good at getting the balance right') was also the maker of another favourite evening dress – an enormous tulle skirt embroidered with irridescent butterflies. 'But my favourite dress of all was never photographed. It was my first Dior dress, white strapless tulle and a vast satin bow at the back. Underneath the huge skirt there was a kind of beehive, fixed like a farthingale. It meant I could move any way, even walk backwards, without tripping up, because of the beehive arrangement.'

In her early twenties, Princess Margaret was mostly dressed by Molyneux, Hartnell and Victor Stiebel, everything designed especially for her, although she did also get a few things from Dior and Jean Dessés. Photographs of those days, often by Cecil Beaton, show a number of glorious, full-blown ball dresses: back-lit, glimmering, romantic. There are also less elegant pictures of the Princess in fur jackets loaned to her by the Queen Mother, which make her laugh. 'How on earth could the papers have called me a leader of fashion?' In the fifties she was 'taken in hand' by Lady Jean Rankin, lady-in-waiting to the Queen. 'It was Jean who introduced me to René, the hairdresser, and also to Simone Mirman, from whom I got my hats for many years.' These days, clothes are acquired from a number of sources. For a foreign tour, she says, there is usually too much to do for one designer, so she goes to several.

'Obviously, abroad, I like to wear English things – Caroline Charles, for instance. Sally Crewe at Sarah Spencer is very kind as she brings me shapes, then does them in special materials. She did a lot for my tour to Denmark last year. And she also made me something I've always longed for, a short evening dress that can also be long. It's made from a scarlet and gold sari, with a separate long underskirt, so I can wear it either way. Very useful.' Roger Brinés, a Frenchman over here, has been designing clothes for the Princess for many years – he produced the azalea dress she wore at the Prince and Princess of Wales's wedding. For really special occasions she sometimes persuades her old friend Carl Toms, the theatrical designer, to think up something for her. 'Carl has me absolutely right. He's used to small people because of designing for ballet dancers.' If she finds she has nothing for private occasions she 'really shops around', going to all sorts of places for things off the peg.

Princess Margaret loves dressing up for evenings, and also for Ascot. 'I get four new things every year for Ascot, and enjoy thinking about them, although I can usually manage to get away without four new evening dresses too.' Materials for the evening she likes to be 'all the richest: velvets, satins, silks – but not much décolletage these days. I think when you're older exposing too much skin is hideous. You have to be very careful not to be mutton dressed as lamb.'

Nowadays, as always, the Princess says she likes London clothes better than country ones. 'I don't ever wear trousers, but skirts and jerseys or shirts in the country, or tweeds with long coats or macks. What I really like best in the day are dresses.' She has a good collection of these, many in fine, printed wools, very simple and comfortable. 'In the evening in private life I'm very unfashionable in that I wear long dresses when most people are wearing short, but it's because I've got so many long ones.' She has firm views about the decline of the old evening ritual of changing. 'People no longer seem to have a sense of occasion. My feeling is that if you're going out in the evening then you want to have a bath and change into clean clothes: you choose something to suit the occasion. Sometimes Sarah appears when I'm changing to go out and says "It's wonderful to come in and find you looking so groomed!" Well, I do think grooming is terribly important. To me smartness is not only to do with clothes, but make-up, hair, bags, jewellery, even nails. They all add up to the finished effect. They're all important.'

The Princess has few definite dislikes. 'I never wear brown, I find it very depressing. I don't much like purple, either, though I like refinements of that colour. I can't bear materials that aren't smooth – bouclé or shaggy tweed – and I never wear silk crêpe. I know a low waist doesn't suit me, and I'm not a very frilly person. I don't really like hats: to me they mean I'm on duty, and anyway I'm convinced I look ridiculous in a hat. If I do wear one it mustn't be heavy – a whole day in a heavy hat is very uncomfortable. I don't like hard hats, either, that you can't put a pin through – I do love feathers, though. For formal occasions, of course, I have to wear hats, often to match whatever else I'm wearing. Many of them come from Graham Smith, who makes for Kangol.'

Apart from brown and purple, she 'goes through the rainbow' of colours. And never a day goes by without her wearing various items from her collection of beautiful jewellery – which includes modern as well as old pieces. 'Well, I'm lucky enough to have some. I don't think I have a favourite stone – rubies, perhaps. I pay great attention to where I put brooches: a wrongly placed one looks awful.

'I'm always conscious of what's in fashion,' she says, 'because without following it too strictly, one must get the line right. For instance, for me, padded shoulders can't be too high. One has to adapt to what suits one. What I really enjoy is seeing how clothes are constructed: finding out how things are made and what of, and I enjoy going to dress shows.' She loves seeing women in pretty clothes, and two whose clothes she much admires are the Duchess of Kent and the Duchess of Buccleuch. 'They're both marvellous coathangers. They could put on any old thing and look wonderful. They have an ability, natural to them as charm, to make anything look good.'

Princess Margaret admits to being a conventional dresser. 'I dress for the public,' she says. 'I have to conform. I care about clothes but I've *never* thought of myself as a leader of fashion.' For many years she was the victim of fashion writers, however, who, generation after generation, pick upon various members of the Royal Family and criticize their way of dressing. Looking back, Princess Margaret sees this trait as a recurring pattern, part of which used to include her, and from which she suffered. 'What they do,' she says, 'is to build up one of us into a so-called leader of fashion, and then knock us down. I remember when they did it to my aunt, the Duchess of Kent, who was wonderfully elegant, in the twenties and thirties. Then it was my turn. They tried to say I was a leader of fashion, which I hotly denied. How could I have been when I was not that smart? And now it's the Princess of Wales. In her interview with Alastair Burnet on television last year she said all the things I was saying twenty-five years ago. Clothes aren't her prime concern. They weren't mine. But the fashion writers persist in treating her, as they did me, as if we were unreal figures from *Dynasty* with nothing better to think about. They criticize if the same things are worn twice, but also criticize if too much money appears to be spent on clothes. So it's hard to get it right, in their eyes.'

Despite criticism, Princess Margaret has continued very much in her own way over the years – in her own words, 'always consistent in a limited fashion, always aware of a sense of occasion. I merely always try to look my best because that's a kind of compliment to those who have invited you to something, or who are around you. It's the least you can do.' As for being complimented herself on her clothes? The Princess laughs merrily. 'Well, when that happens it's very nice. It gives one confidence.'

Atalanta Madden

FARMER'S WIFE

'My clothes are influenced by what I'm reading – Jane Austen, and I turn nineteenth century . . . The only reason I care about them is because I find the world so beautiful I don't want to be a blot on the landscape.'

Everyday clothes for working life on the farm. The brown tweed coat was given to Atalanta by her sister, the shooting hat belongs to her husband.

Atalanta Madden is the youngest of the three famously beautiful Clifford sisters, much photographed in their youth by Cecil Beaton. Now married to Michael Madden, they live on a Cornish dairy farm where they also breed racehorses. Five miles from the nearest town, Atalanta is an active farmer's wife, helping with the milking, the calves, driving the tractor and riding horses out between seasons, apart from coping with all the domestic side of life. The Maddens go racing all over the British Isles in summer, and have one good holiday abroad a year, but rarely go to London. Atalanta has two sons and a grown-up daughter.

'It was all a disaster in the beginning,' she says. 'Being the youngest of three daughters my mother gave up with me and anyway I disregarded her advice. I remember going to Ascot for the first time in the fifties, wearing a sleeveless sundress ... A guest in the box wrapped a fur round my shoulders which I sweated in for the rest of the day. That about sums up my start in clothes. I was always wearing handed-down things and finding them a frightful embarrassment.

'At twenty I married Richard Fairey, and my sister Pandora said, "You're now married to a rich man. Your clothes won't do. I'm taking you to Harrods." Richard bought my first handbag – he thought I looked so scruffy, and Pandora dressed me. She did very well.' (Atalanta met Richard Fairey when she was learning to fly. When I interviewed her in 1959 she was wearing the ultimate in chic flying clothes – a white leather boilersuit to match her own white single-engine plane.) 'Richard took me to Balenciaga, so I had some lovely things, but I was *inconsistent*. I remember being very dissatisfied: my bosom wasn't big enough, and so on. Nor for ages did I find out what I *liked*.'

In the sixties, after the death of Richard Fairey, Atalanta married a bloodstock agent. 'It was probably not till then, the early sixties, when we had to mix with a rich, fashion-conscious set of racing people, that I began to take my appearance seriously. I tried to comply with the uniform but make it a bit different. The racing uniform for women in those days was a tweed Chanel suit or a Hardy Amies skirt and jacket with a silk shirt. Then came the velvet jacket, Russell & Bromley or Gucci shoes with a gold band round the heel. I couldn't ever face the velvet jacket but I did buy some tweed suits – either very Victorian, or very nineteen-twenties, slightly different. I was probably always badly dressed and that still applies today. But in the sixties at least it became more enjoyable.

'Now, very, very rarely, dressing up is an enormous pleasure. The last time I had a ball I went to London to the sales with Pandora. We had great fun buying identical dresses for half-price – two middle-aged old bags flapping our legs in a Charleston in front of the dressing-room mirror. The young salesgirls' eyes were on stalks. They simply couldn't decide how old we were.

'I can't buy clothes in Cornwall,' Atalanta goes on, 'so Laura Ashley mail order is my mainstay. I go for a shopping spree in London perhaps once a year, otherwise my daughter, Leanda, who has wonderful clothes, gives me things. I get shirts and jeans for milking in Wadebridge, and jerseys from my son Henry's second-hand clothes shop at school. I keep certain things for ages – others, terrible mistakes, don't last five minutes.

'What I should like is to look distinctive without being a joke. I won't wear sweeping skirts while others are in minis. I don't follow fashion at all, though do vaguely know when minis go in or out: I wear what I like. Horrors to me are trouser suits in any form, jersey materials, acid green, flamingo pink and purple. I don't like synthetics and only wear them for milking. Frills I don't much like, though a few are okay by Laura Ashley, and I don't suppose I'm much of a lace person. I've never been too keen on Monsoon clothes and I've had to give up chrome yellow as Michael doesn't like it – he only notices what he doesn't like. I dress very much for Michael. His mother is always beautifully dressed, so I try to be like her though I never quite manage it. I hate tights unless they're coloured – I like stockings and suspenders and very feminine underclothes.

'I love cotton, even brushed cotton for winter, and I love waists as I have a small one. Blue is the colour I find least painful to a worn face: as Oscar Wilde said: "A girl is always safe in blue." I like to blend with the background, I don't want to stand out. Not much dressing up goes on in Cornwall. On the farm I'm always in jeans and jerseys and dressing every day is a matter of three or four layers – we've no central heating and life revolves around the Aga. There are lots of small dinner-parties with friends nearby – poker and Trivial Pursuits. And for them my evening clothes are really other people's day clothes. Everyone's houses are reasonably chilly and that's where Laura Ashley comes into her own: wonderful blue tweed baggy trousers, and a navy and red needlecord striped dress, both of which I've had for ages. I wear one of them. I've also some nice things from Laura Ashley I can mix up – a velvet coat and skirt, and black velvet jodhpurs. For shooting I wear one of my son's school suits chopped up, and for racing, though I don't much like hats, I wear a real nineteen-twenties cloche. I've few things really to dress up in – an old white satin bomber suit, and a new nineteen-twenties electric-blue chiffon evening dress from Harrods. But going to weddings is the one time I have fun dressing up. I've two Dior hats from my mother-in-law, both thirty years old. I wear one of those and really enjoy putting it all together, but that sort of occasion is pretty rare. I *don't* spend much money on clothes as I have expensive tastes in holidays, and would rather spend the money on them.'

Shoes, Atalanta finds, in common with most others, are very difficult. 'Mostly I buy them in Wadebridge, second-hand for £3.50, sometimes beautifully made old ones. I have one or two pairs of good ones that last for ages, but as for my evening shoes, they're all chewed up by the dogs . . . I used to have a lot of jewellery, married to Richard, and I loved it. He bought me staggering things but ninety per cent of it has been stolen or sold. I liked the diamond days but don't miss them at all: owning jewellery didn't mean anything to me. I just have some lovely pearls, now, some amber, and junk jewellery, and a semi-precious necklace we bought when a filly won the Cherry Hinton at Newmarket. It's made of round agates, that look like cherry stones, and crystals.

'I'm not a great observer of other people's clothes,' Atalanta goes on, 'though I do get complimented myself, and like that. I think the British rag-trade is the best in the world at the moment, and the young upper-class English girl can be the most elegant in the world: in fact, I think the young English as a whole are pretty good at dressing. They use clothes available in a dramatic and individual way. But the middle-aged, sadly, look very dreary. Clothes don't seem to be made for them here, though they are in Paris. Only Laura Ashley and *haute couture* designers bother with them. Generally, clothes in London seem to me neither practical nor suitable – no wonder there's so much mutton dressed as lamb. I definitely think it's harder to dress well in middle age now than it used to be. Lady Beit is someone I admire very much – she's a very elegant lady. I'd like to dress like her at her age. And Princess Alexandra is always well turned out. Then I'm an enormous admirer of the Princess of Wales – she dresses with great dignity and flair.'

Atalanta's own form of dressing she describes as 'unquestionably romantic, but non-conformist'. In a way that is somehow traditional among beautiful, middle-aged Englishwomen, she looks marvellous (and extraordinarily young) by day in jeans and no make-up, and then is transformed into a great beauty at night, however simple or underdressed her evening clothes.

'I believe clothes are an illustration of one's personality,' she says, 'and I'm very influenced by what I'm reading. Jane Austen, and I turn nineteeth century. *Le Misanthrope* is reflected in the nineteen-twenties clothes I bought a few years ago, and I was totally nineteen-forties after seeing *The Jewel in the Crown*. The only reason I care about clothes is because I find the world so beautiful I don't want to be a blot on the landscape. There are such marvellous views from our house – nobody would want to see something in an acid-green trouser suit walking across them.'

Ruby satin cocktail dress, circa 1920s, hand embroidered. 'Probably one of my mother's,' says Atalanta, 'the sort of thing I'd wear out to dinner in London but it's unlikely there'd be an occasion for it in Cornwall.' Typical of Atalanta's original choice of colour is purple tights with blue shoes.

Frances Maloney

BUTLIN'S GLAMOROUS GRANDMOTHER

'I was the first in Leigh to wear the New Look – a big picture hat, ankle-strap shoes, pink rose in lapel and hat, head in the air . . . everyone stared as I walked by.'

One of Frances's own designs, made by a dressmaker. 'Pure wool,' 50p a yard up the market. The dress is A-line with a red satin bow to match the lining of the cape. The hat came from the market, too. I just added a band of red to tie in. It's my favourite outfit.'

Frances Maloney was born in the mill town of Leigh (now a mining town) near Manchester. Her father was a miner and died in his thirties. Her mother took in washing to support the five children. Frances was married, briefly, at twenty and had a son. At sixty-five she retired from her job as a clerical officer in local government and still lives in Leigh, where she shares a house with her sister. For the past twenty-five years she has entered Butlin's Glamorous Grandmother of Great Britain Competition and has reached the finals seventeen times. In 1984 she came third, out of ten thousand. This year will the last time she enters – 'enough is enough', she says. A local celebrity and written about in Butlin's Golden Jubilee Book, Frances spends considerable time organizing every detail of her competition clothes, and she still does part-time work modelling as the bride's mother for a bridal shop in Earlstown.

Frances Maloney

'At first there was rivalry between my two aunts as to who could dress me best,' says Frances, 'then they emigrated to Australia and my mother felt it hard. We'd been spoiled, now there was nothing. In fact, my school teacher kept me clothed – that's how badly off we were. But from a very young age we saw nice things because of my mother taking in washing from the big house. Even when I was very young I vowed and declared that one day I would have nice things, too.

'I was in a good gang of girls at school, and will never forget our first ball dress. I say "our" because we all had the same. I found this pillar-box red stuff, sixpence a yard in the market, and I designed it. All the girls wanted the same style – one of them made up the dresses. The only trouble was we hadn't any money over for rouge. So you know what we did? We crept into the church, spat on our fingers, and rubbed them over the red covers of the prayer books. I did all the girls' hair. I remember the smell of cheap Californian Poppy scent which we put on to take away the smell of burnt hair. I designed white satin blouses for the gang, too: I've seen the styles coming back, lately. Yes, from a very young age I was interested in clothes. I always had an idea of the finished product even though I couldn't sew. I always arranged to look nice, never ridiculous, even when I had nothing.'

During the war, Frances worked as a bus conductress. She smiles at the memory. 'Well, I was told off because I narrowed the trousers and made the jacket into a safari jacket! I was quite the best dressed conductress.' Having 'accumulated a bit of money', she was able to start dressing in the way she had always imagined. 'There are more things in life than clothes,' she says, 'and I don't make them my priority, but I do like them. If I plan right I find things don't date – I just alter accessories, or add belts and bows. A lot of modern styles I wouldn't be found dead in – the English lady isn't suited to all these gimmicks. I dress for myself and know my pattern. I think we are all *seasons*, and I know I'm autumn: I would never try to be spring. I'm a typical English lady – pear-shaped, busty – I always wear a good foundation, I hate women of my age with no bras. A *suit* person is me, down to the ground: in a good suit I feel a million dollars, and I love capes, too. I like shirtwaisters, and basic tailored things with long sleeves, and I love *bows* – they're very feminine. I wear skirts and shirts but they must have a cardigan. A skirt and shirt on its own doesn't flatter: you don't feel dressed. I detest strapless evening gowns at my age, I like to cover my neck. I'd never buy an evening gown with gold – I go for silver, the sparkle. I wear trousers sometimes, but never for a special occasion. My hair is my hat.'

Frances has always 'had an eye for a bargain' and enjoys scouring junk shops and antique shops where she finds good buttons and beads, and jewels for trimmings ('I like trimmings to be used discreetly'). She still shops in the market for both materials and clothes. 'I like to look for a Made in England label,' she says. 'I've a dressmaker in Leigh, and have about fifty per cent of my things made. I do go to shops in Manchester and Bolton, but there's so much duplication today. There are loads of boutiques filled with a lot of rubbish: I like something individual. I go to a place and the salesgirl will say, "Now, this will suit you". "No, it won't," I say – I know what suits me before I try it on.'

For her competition dresses Frances used to love velvet – indeed, she wore black velvet on the occasion she came third. 'But then I stopped,' she says, 'because you can't get the quality any more. Now I'm fond of silk jersey. For this year's competition, I've chosen a black silk jersey with a cummerbund of cerise sequins, and sequins on the *Dynasty*-style padded shoulder. I like satin, if you can get the real thing, and real wool. I don't mind synthetics if they're good and soft, but I don't like Crimplene. My favourite colours are blues, pinks, blacks and reds. I have a red suit and people say "Wherever she goes in that she's alive!" I feel bright in red and pink, sexy in black. I don't like white, beige, brown, grey, yellow and gold. I'm not a gold person at all. Gold and fur coats leave me cold. But I like shoes, Italian ones best. Good British shoes are hard to find these days – they're a rubbishy lot. What I'm after is slingbacks with high heels, which go with longer length skirts. If I was younger I'd fancy those

Frances Maloney

romantic *Gone with the Wind* clothes. As for tights, I never seem to be able to find the right colour so I dye my own with artists' paint. It's cheaper than dye, and you have to be thrifty. I don't have much jewellery, and nothing real. I keep a look-out in the market and antique stalls and just wear the occasional brooch, and earrings for competitions.

Frances doesn't think much of the way Englishwomen dress today. 'It was good after the war,' she says. 'I was the first in Leigh to wear the New Look – big picture hat, ankle-strap shoes, pink rose in my lapel and hat, head in the air . . . everyone stared as I walked by. The *young* today don't make the most of themselves. They've so much more money and opportunity than I ever had. If I could have had what they have . . . I think they should take a leaf out of the Princess of Wales's book: and I admire the way the Queen Mother is always turned out – she's a lady in every sense – and the Duchess of Kent has really come into her own.'

Incredible effort goes into each one of Frances' dresses for the Glamorous Grandmother Competition. When she first entered, twenty-five years ago in Skegness, she wore a black velvet dress and 'a corsage of blue flowers and mother of pearl, with a matching band in my hair'. She went through to the finals, and, every year since, there have been weeks of careful planning for the next event. But once a dress has been in a competition it has little active life left, though it may be worn at a few openings. 'Actually, competition dresses are very wasteful as they can't be worn except the once or twice,' says Frances, 'but I can't bring myself to throw them away: they're my memories. I look at one and remember just how it all was. I live on the memories.' Among the old competition dresses that crowd Frances' wardrobes, thriving on careful treatment

This pure wool suit with leather trimming was a 'showstopper', Frances says, when she modelled it at a bridal shop in Earlstown. The bride's mother was the part she played; she acquired the suit with its pouch-back jacket for a nice discount.

like well-preserved stars, is a ruby velvet number, designed by Frances herself, that made even Norman Hartnell gasp in 1978. 'He said it was incredible, he'd never seen anything like it,' says Frances. Another dress is richly trimmed with feathers, which she loves, though she claims the good ones are hard to find these days.

The production of a competition dress involves very hard work. Having decided on a style, Frances goes in search of the material. The 'shocking-rose pink pretend silk' for the 1985 dress cost £2 a metre in the local market. It was made with a drop waist – 'very flattering to a pear-shaped lady' – and batwing sleeves, embroidered with trimming from another dress with loose beads (from the market) added. Satin shoes were dyed to match the dress, and decorated with pearls. Tights were dyed with paint to match, also. Homemade earrings were made to match the trimming. Angela, Frances' friend and supporter, who goes with her to all the competitions, made a bag from a piece of the pink silk with yet more matching trim. Attention to such detail is vital to each competition outfit. It all takes time and much trouble, but is worth it for the excitement and enjoyment.

'I make a lot of effort with my clothes every single day in fact,' says Frances. 'I always take trouble with my appearance, and every Thursday I have a pampering day, doing eyebrows and nails and hair and everything. But the appreciation I get makes it worthwhile. You'd be surprised by how many grannies ask my advice. I'd say I was a *show-stopper* of a dresser, an individual. I like to be noticed, and as long as I feel Queen of the May I don't care what anyone else thinks. I do cause a lot of comment, which I like, seeing as it was so hard when I started.' She smiles. 'Actually . . . I know just how the Queen must feel when she's going off all dressed up for an opening night, because when *I'm* going out, dressed up, people come and stand at the door and watch me going off in a taxi. It's very rewarding.' And her friend Angela can vouch for the fame Frances has acquired through her show-stopping clothes. 'It's true,' she declares, having witnessed for years the impact made by the glamorous grandmother, 'Frances is *the* legend.'

Martha Fiennes

STUDENT OF FILM AND TELEVISION

'I wear a hat for dinner parties ... I flirt with being an outrageous dresser, but would never do anything vulgar. If the outrageous is there, it's in the detail.'

Long angora cardigan, Indian shirt and black leggings – very much everyday wear for Martha, though she wouldn't wear the hat at college. That she made herself from the collar of an Oxfam Grannie's coat.

Martha Fiennes is in her third and final year at Harrow College of Higher Education, where she is studying for a degree in film and television. She lives with friends in Notting Hill Gate and tries to survive on an ILEA grant. To supplement this, she takes part-time jobs in various clothes shops. She drives a 'clapped-out' Morris Minor and spends occasional weekends with her large family in Wiltshire.

The Englishwoman's Wardrobe

'Clothes,' says Martha, 'are my biggest weakness, my financial priority. I don't know what I spend on them a year but it can't be much because I haven't got much. But I make a lot of my own things, which of course is a great saving. I've never learnt and I'm not brilliant, but I can make something in an afternoon, cut it out without a pattern and know it's exactly right for *me* – I can't make for others. My interest in clothes was awakened when I was about sixteen, doing A-levels. I became obsessed by them – pathetically obsessive, really. Today the obsession has toned down, though every single day I give considerable thought to what I'm going to wear, and if I'm going away for the weekend I think: "What shall I wear on Sunday? Should it be different from Friday night?" If I get up in the morning feeling aggressive and confident, then I'll put on bright colours. If I'm feeling frail from the night before, then it will be black from head to toe. I might change four times a day if I don't feel it's right. In a way, it's distasteful, that attitude, such preoccupation, but that's how I feel. I'm observant of other people's clothes, though not critical of my contemporaries.

'I'm very conscious of what I've got in my wardrobe and how well kitted out I am. I buy mostly from markets, the Portobello Road and Camden – I love the atmosphere there. Looking, discovering, buying – it's all part of the process of creating: then you find something and it's a complete thrill. You wear it in a way no one will see in a magazine or a shop, that's the thrill. I've got an eye

'You're allowed the odd eccentricity,' is Martha's view, but once wearing this old hat to a dinner party somebody remarked she looked frightening. The black cotton dress hung with huge swatches of silk came from Kensington Market – Martha added more silk.

SARAH
beloved wife
and mother

for a bargain, though not always the bargain I was looking for. I go off intending to buy a jumper and come back with three Victorian nightdresses. Apart from markets, I go to Miss Selfridge, which is brilliant for foundation things – by that I mean black cotton leggings, which I live in. And they have a marvellous black T-shirt stuff skirt, £9.99, which masses of my friends have bought: you can dress it up any way. I go to Marks & Spencer for underclothes, but if I'm in a spendthrift mood then I go to Joseph in South Molton Street. I'm quite practical about my wardrobe. Even if I love something I always think, have I got anything to go with it? I suppose I buy *something* about every one or two weeks, but I keep old favourites year in, year out. With most things I can tell immediately whether they're going to serve their purpose, and I am always right. Other things – I can't tell which way they're going to go. I give things away when I'm fed up with them, as I hate having clothes in my cupboard that I don't like. I can't afford to make many mistakes.'

The kind of dresser Martha aims to be is one who has learned the art of *balance*. 'I like a baggy skirt with a short jumper, or a straight skirt with a huge jumper. I never wear dresses in the day, but *detailed* layers. I love long cardigans which I sometimes knot at the hem. I'm *slightly* influenced by what's *à la mode*, but it won't be the letter of the law, more the spirit of the law. For instance, if I saw something in Portobello Road, say a very feminine Victorian shirt, I would be as confident in that as in a fashionable black polo neck. The most important thing to me is that clothes *work*: they must have balance, and a sting in the tail – earrings, make-up, long puffed sleeves, pearls in the hair, whatever – if they work, that's more important than making a huge statement saying *Look at me*. The *way* you wear things, carry them off, is so important. I didn't consciously create my own stamp, but people say I have one. I flirt with being an outrageous dresser, but would never do anything vulgar – not sexy or obvious – more, unusual. If the outrageous is there, it's in the detail.'

In common with many of her generation, Martha is extraordinarily fond of black. 'It's addictive with me,' she says. 'It makes me feel confident. It's the minimalism of it. That's where its strength lies. Something often nags me, saying I must get away from it: it's worn so much it doesn't lend you much identity. But I love it. I *do* like other colours, too: that green-baize-door green, rich deep purple, cool colours rather than warm. I love just plain white in summer, but I hate pastels, and dirty blue, and that pink with purple in it. I never wear yellow or orange, and I simply can't wear turquoise, though I like the idea of it being like Hockney swimming pools. I'd love to wear pale grey, but I can't. The materials I like are real wool, everything from cashmere to coarse fisherman-y, real silk – crushed to Thai, and real velvet I adore. Also fine quality cotton that hangs well. And linen. I love it looking crushed. East End markets are brilliant for materials, so is Borovicks in Berwick Street. It has wonderful theatrical materials, fantastic gaudy organza stuff. Peter Jones is good for standard things. I hate corduroy because it's stiff and has Sloane Ranger associations, and all synthetics, even mixes. Wool jersey is awful and brushed cotton – well, that's obvious. As for track-suit tops and bottoms in pastels, they're the epitome of ghastliness. Anything acrylic is anathema to me . . . that awful feeling.'

Some years ago Martha worked as a sales assistant in Belinda Bellville. 'People would come in and say they wanted a cocktail dress or a ball dress or a wedding dress – the range is vast. I would look at them and show them three extremes, thinking they might like one. They'd look at me and laugh scornfully, as if I was ridiculous not to *know* exactly what they had in mind. . . . For myself, I don't feel right in anything tucked in at the waist. I like low-slung belts. I don't wear high necks or frilly collars, though I do like frilled cuffs and sleeves. I love the Indian influence – those lovely simple necks and leggings, which aren't meant to look like tights. I don't go for stiletto or decorative shoes, and I obviously hate anything strappy. I wear black suede usually, with low heels. I could say a lot about shoes. Many of them are *so* badly made. However, men's shoes have changed my life. Since I've been wearing them, I realize what a fantastic time men

Martha Fiennes

have: their shoes are comfortable and warm and the old ones I find are beautifully made. I love the patent evening ones with a bow: I get them second-hand from a friend. Shoes aren't a great priority, but if I had a lot of money I would go to Ad Hoc and Emma Hope and Bertie.

'But *hats*,' Martha goes on, 'hats I love to pieces. Dark, wide-brimmed crumply straw in summer – I put scarves round them like Isadora Duncan. Or berets, or extraordinary hats that shoot out – usually old again. I get them in markets for five or six pounds. I'd never spend thirty-five pounds at Hyper Hyper. Hats have only recently become accepted among people of my age. They used to be thought pseudish. I wear them for dinner parties.'

Martha loves dressing up in the evening, and it is then she puts on a large selection of her bits and pieces, apart from her wonderful hats, for which she is renowned among her friends. 'Jewellery is the icing on the cake for me. I haven't that much that is real, but a great-aunt left me some fantastic pearl chandelier earrings. But my speciality seems to be finding things in junk shops – buckles, buttons, brooches, pins for the hair and hat pins.' She carefully festoons herself with her junk-shop prizes and the result is arresting, original. 'I dress a bit for myself, a bit for everyone. I enjoy the process of putting it all together, trying to evoke a mood, an atmosphere, a style. Although I'm a bit influenced by what's going on, I always like to retain a feeling of my own stamp. In my experience, a lot of the young are very well dressed in London, though there is a *clone* element, which I find depressing, ninety-nine girls all alike. I'm fascinated by the way middle-aged women dress. . . . the only one I can think of whose clothes I admire is Belinda Bellville. There's a lot of room for improvement in Englishwomen's clothes, but I don't think it's a matter of money. It's a lack of interest, not knowing how to present themselves. It's so easy to be versatile, adding and taking away, but Englishwomen just don't seem to have the imagination. On the whole, they don't have the talent for putting things together and making themselves pleasing to the eye.'

Previous page: 'Working, clothes come definitely second. They have to be completely practical.' This is someone's coat she found hanging up, with Martha's own riding boots and colourful scarves.

WENDY BALL

WIFE OF THE WARDEN OF KEBLE COLLEGE, OXFORD

'Oxford is not well dressed. It was much better ten or fifteen years ago. Now, I admit, I do find myself trying to hoick up the general standard.'

White silk coat-dress from Wallis, five or six years old, whose shoulders were 'very extreme' when it was new. An adaptable favourite, Mrs Ball wears it for 'weddings,Encaenia and degree days at Polys'.

Wendy Ball is married to Christopher Ball, Warden of Keble College, Oxford. As mother of six children (three still at home) and wife of a Head of College, her life is active: there are constant dinners and entertaining for her husband's colleagues and college guests, most of which Mrs Ball organizes and cooks herself. All first-year undergraduates are invited to lunch in the lodgings, which means one large lunch party a week, every week, in term. As a 'part-time' job (it takes a pretty large part), Mrs Ball has become an organizer of international conferences held in Oxford. In any free moments she runs a small cottage industry: an expert needlewoman, she makes one-off clothes for other people – very elaborate things, sometimes, such as the wedding dress last year for a bride married at St Margaret's, Westminster. She makes fifty per cent of her own

Wendy Ball

clothes, and is one of the few Oxford wives who seems to have any interest in clothes.

'I live on a bike,' she says, 'so I have to have clothes that aren't going to be ruined. But I'm not really a trousers or jeans person – I'm not a good trouser shape. I do have one pair of white trousers from Marks & Spencer, mildly snazzy, but they're not an everyday thing for me. Although my life is hectic, I think it's very important to make the best of oneself every day – you owe it to those around you to try to look as nice as you can. I hardly ever wear make-up in the day as I like to make a change, an impact, by putting it on at night. But I do love variety. Because of work, gardening, and going out or entertaining in the evening, I often find myself changing several times a day.

'I'm not at all interested in what's currently in fashion unless I happen to like it or want to adapt something for myself. And I'm not interested in not being like anyone else. That doesn't worry me. I wear what I want to wear.' And what Mrs Ball wants to wear is certainly a sartorial improvement on the general standard of women's clothes in Hall.

'At college dinners, on guest nights with black ties, I feel at liberty to wear a long dress, or, if it's short, it's elegant, probably silk. I try not to spend a *great* deal of time thinking about clothes – I'm too busy. I certainly don't plan my wardrobe like women are advised to – a few basic inter-matching colours and so on. How dull that would be. I would hate to think I couldn't suddenly buy something yellow because that would upset my whole colour scheme – that would be very upsetting.' She produced an impulse buy of a canary-yellow sundress from Marks & Spencer, which it would have been sad to deny herself. 'Rather nice, don't you think? A lot of swing!'

'Clothes for me,' she says, 'are a sort of on-going thing. If I like something – and on the whole I don't buy unless I am very sure, so I don't make many mistakes – I never throw it away until it's completely worn out. I love evening things best. One of my favourite dresses, an Ossie Clark long crêpe evening dress, is thirteen years old and still going strong. I always try to buy or make things that are going to work for me for twenty years, things that won't date. I'm not a classic dresser in the strict sense, but because I'm tall things tend to look quite grand. The idea of a grey flannel suit, or skirts and shirts, bores me. Though I do have one suit, a wine suede, quite old, which I call my "going to schools to give away prizes suit". I love cloaks rather than coats, and make them for myself. On the whole I'm a relaxed dresser, I would say.'

When it comes to designing and making things for herself, Mrs Ball begins by finding the material. 'I get a lot from Liberty's sales. Once I've found what I want I do a few drawings, usually adapting some pattern or design I've found. Then I sew away listening to Radio Three, and really enjoy it.' She showed me a simple, long-sleeved white smock, illustrating her original choices of material – it was made from a damask tablecloth. Most of her home-made things are distinctive for their use of small buttons, edgings and trimmings. 'I do think piping and buttons add a finish, make things more individual,' she says. 'I have to admit I'm quite influenced by Annabelinda in that way. I loved her things when she started . . .' She has several Annabelinda originals, including a much-loved eight-year-old long Indian gold silk skirt and jacket, both edged in brown velvet. Another, also Indian silk, was a dramatic print of exotic birds.

Mrs Ball's most admired designers are Jean Muir, Caroline Charles and Benny Ong. 'Then I loved Ossie Clark and those heavenly Celia Birtwell prints. And I think Zandra Rhode's *fabrics* are out of this world. I'd give my eyes for the one in the textile room of the V and A. I've never had a real Jean Muir.

In Oxford the shops she likes best are Wallis, Tyrwhitts, Blue Stocking, and Marks & Spencer for underclothes and 'the occasional shirt'. She produced a white synthetic silk one that looked very Jasper Conran. On the whole she does not like synthetics, though 'they have their uses' and are sometimes, as her black printed viscose cocktail dress and jacket by Jinty prove, pretty convincing crêpe de chine. A favourite coat bought at Wallis five years ago is a cream raw silk with dramatic shoulders. 'I wavered about buying it at the time: there weren't many such shoulders about then. But it's been a great success. I wear it with a little white felt *Aida*-shaped hat, which always amuses people and cheers me up. But then I love hats. I have about eight which I wear for special occasions. I think the point about hats is that they should be fun, make people smile, so that you enjoy wearing them. One of the most amusing ones I made myself.' It was a yellow silk Jackie Kennedy pillbox with a puff of yellow net.

What Mrs Ball does not like, on herself, are frills. 'I like clear strong lines and clear colours. I suppose most of my things are black and white and cream. I can't take a dropped waist any more, and I always wear things my own length. They've got longer since I've got older. I like a variety of styles.

'I would spend a fortune on jewellery if I could,' she goes on. 'I like to encourage young designers. Wally Gilbert, whose things I buy at the Oxford Gallery, I'm particularly fond of.' She brought out some beautiful earrings in the shape of lace wings, made of silver and gold filigree. Breon O'Casey, who makes strong, simple silver and gold things is another favourite. 'I like Victorian cameos – things from any period, really, as long as they're not itsy-bitsy or little modern diamonds.'

Shoes hold no particular interest: they have to be practical with all the bicycling and she feels if she could spend a minimum of £60 a pair she could probably find nice ones in London. 'I'd like to buy a

Mrs Ball declares she 'lives on a bike', but she takes care not to look as unattractive as most Oxford cyclists. This jacket and skirt she made herself from Liberty cotton, adapting a pattern.

lot from Bertie. I had a favourite pair of Mary Quant's from the early sixties, which I'm still wearing. Most of my tights are black or white, sometimes maroon or other colours to go with what I'm wearing in the winter. Never flesh-coloured.'

In general, Mrs Ball feels Englishwomen's clothes are 'very boring. They're still so tentative about colour. Walking down any street most of the women you see are extraordinarily drab. Except for the young – I love some of their clothes. I wish they'd been like that when I was young, instead of all those little white collars and gloves. But the good thing about clothes today is the freedom to do exactly what you like. I can't think why more people don't take advantage of it. After all, clothes tell a lot about character. I think if you're interested in them, then you shouldn't hide that interest. You should make a bold, extrovert statement. You should expect attention. When I put on an evening dress I don't want it to go unnoticed ....'

British reserve forbids extravagant comment, but when her clothes do elicit admiration Mrs Ball is well pleased. Her husband, she says, has quite a good eye, but she trusts her own judgement more than his. 'I wouldn't wear something he really hated, but on the whole he likes the things I make and choose. Oxford,' she adds a little wistfully, 'is not well dressed. It was much better ten or fifteen years ago. Now, I admit, I do find myself trying to hoick up the general standard.'

Previous page: Made from a Jean Muir paper pattern, now discontinued, this black crêpe evening dress is one of Mrs Ball's favourites, much worn over its six years of life at dinner parties and formal occasions.

JILLY COOPER

NOVELIST AND JOURNALIST

'I would definitely describe myself as a sexy dresser. That's what I dress for, that's what it's all about, to be sexy.'

Suitable clothes for walking the dogs are a priority in Jilly's life; this coat was one of the perks she sometimes receives from television – she wore it in an advertisement.

Jilly Cooper and her husband, Leo, live with their two children and various dogs in a beautiful house in Gloucestershire. Jilly, who has become an avid lover of country life having given up her (Barnes) Common years, tries not to go to London more than once a month. She works extremely hard writing, for ever goaded by constant deadlines. The pattern is sometimes broken by appearances on television, and there is a certain amount of Gloucestershire entertaining at weekends. Also, she walks the dogs twice a day.

'As a writer,' says Jilly, 'I'm supposed to look odd.' She sits in a delightful, untidy bedroom in which the vast dog baskets are apparently of far more importance than the shelves of clothes. 'I've always been badly dressed,' she goes on to admit, 'because in order to be well dressed you have to spend a lot of time thinking about your clothes. I haven't got that time, and even if I had I probably wouldn't bother. Food and wine and clothes – to me they all come into the same category – are all very well, but they're secondary.

'When I was young,' she explains, 'my mother insisted on putting me into tweed suits. So as soon as I grew up I rebelled and went straight for man-made fibres with plunging necklines. I love man-made fibres! They cling so well – Viyella and cotton don't cling. I've always had a big bum – my mother's solution was to put me in clothes that don't *touch* – mid-calf straightish skirts. My revenge at twenty-one was those really lurid synthetics. I loved those vamp dresses at night – in fact, I still do. Then I wrote a book about class and discovered that no one should ever wear man-made fibres and anyway Leo, who has very good taste, would throw a moody when I wore them. But, risen socially, I daresay I've lost some sex appeal . . . .'

'I've also been a bit schizo about clothes – two distinct types, really. In the daytime, working, I live in skirts and shirts and jeans and jerseys and, since my son has been going to school, prep school things. I wear a lot of his trousers, cricketing shirts and braces. Then, in the evening, if there's a party, there's all the ritual of dressing up. I love that. I'm very vain in that I'd hate to look awful at a party. That's probably a very British attitude – I mean, on the whole Englishwomen look best at sporting occasions (Barbours are very sexy, I think) and at night. They're not very elegant in general, but they come into their own on special occasions like the Fourth of June and Badminton and upper-class weddings. And then they look wonderful, their best, at night – Scotch mist shoulders rising out of pretty dresses, and all that beautiful skin. But English *men* aren't very interested in what their *wives* wear, on the whole, are they? All those women trying hard to look good at parties are actually gritting their teeth against infidelities. I could count on one hand the men I know who notice and praise their wives' appearances . . . not very helpful. Leo, actually, *does* notice, acutely: I dress a bit for him. The children are passionately dismissive about my clothes, and very conservative. But they dress very well themselves, never having been given any direction.

'I should probably take Leo with me when I go shopping for clothes because I do make dreadful mistakes – my cupboard's full of them. They're usually prompted by an occasion: I never think of clothes except in terms of what I'm going to do – go to the Tory Party Conference, appear on television, whatever. (I once wore braces on *What's My Line?* That didn't go down at all well.) I never go searching for something with a preconceived idea. I've no courage of my convictions, I just see something I like and buy it spontaneously. That's often fatal. There was the occasion of my son Felix's confirmation. I bought this blue suit from Jaeger. It was quite the wrong blue and anyhow ten other mothers were wearing it – complete disaster. Another disaster was

Right: For everyday at home in Gloucestershire, Jilly's clothes are often courtesy of her son Felix – striped shirts and school braces.

Overleaf: Interesting ratio of dog space to wardrobe space . . . Jilly's love of dressing up at night does not necessarily mean in exotic or expensive things. This 'favourite dress of all time' was bought from Miss Selfridge in the late sixties for £8: its success is perennial.

when I bought a five-hundred-pound white silk ball dress from Brown's in Dublin, visualizing a life of parties which didn't happen. I never wore it. Leo buys me lovely shirts, though in hopeless colours for me, and antique jewellery which I love because it's from him, but I wouldn't mind if I didn't own any jewellery. My mother gives me presents of beautiful cashmere jerseys – I don't much like cardigans.

'The thing is,' says Jilly, 'months of my life go by without giving clothes a thought, then, from time to time, I think about them with deep passion. I certainly don't plan my wardrobe in the practical way that you're supposed to. When I was married in the early sixties I bought clothes all the time. I didn't have much money but I still spent a lot on clothes, although even then I didn't think about them in the same serious way as some people. For instance, there was a girl in the office where I worked, married to an Italian, who actually spent a great deal of time *thinking about bags*. Have you ever heard anything so ludicrous? Accessories don't matter a stuff. In the late sixties I did manage to buy two incredible classics that have gone on for ever since. One was a black velvet cloak, which was the best buy ever, and the other was the sexiest black dress I've ever had – my favourite of all time, long sleeves, plunging neck. It always works. It came from a sale at Miss Selfridge – eight pounds. I often wear it when we go to dinner parties round here when you can often get caught out. They say it's a kitchen supper, then you go into the dining-room. But my eight-pound dress is always a success.'

Jilly has very definite ideas about what kind of dressing she does *not* like. 'I'm allergic to dons' wife dressing,' she says. 'Middle-aged women in high-necked frills. In fact, I don't like frills at all, especially for myself. I *hate* wide shoulders, bright colours and capes. Parkas I think are absolutely revolting, as are trousers with straps under the foot. I also hate hats – Leo says women in hats want to be screwed. Once, reporting Princess Alexandra's wedding for the *Sunday Times*, I had to wear one, but it looked awful. I don't think baggy trousers are very pretty, though they're lovely and comfortable. I don't ever wear trousers in the evening. Dressing is very tricky at our age,' she goes on. 'You can easily descend into the frumpy and give up, though that's not necessary. I'm passionately interested in what's in fashion, though I have little access to it. I may miss jodhpurs one year and catch up the next. But I won't wear something currently fashionable if it plainly doesn't suit me: in any case, Leo would forcibly stop me. Also, if I have a bad evening in a dress, or Leo really didn't like it, then I'd never wear it again.'

Since giving up man-made fibres, Jilly's favourite materials are now silk and cashmere. 'I'd like to live in them,' she says, and her large collection of 'millions' of silk shirts – some of them very old – enable her to do this much of the time. She also loves T-shirts, which she wears with skirts most of the summer. Favourite colours are black, cream, beige and most pastels. 'I love pale pink, blue and aquamarine. If you have a red face you can't wear anything red or tan. I always like to be underdressed rather than overdressed,' she adds. 'The clothes I wear in London are no different from my country ones.'

When she lived in London, Jilly used to buy a lot from Jigsaw in Putney, and she was 'very drawn' to Miss Selfridge, Chelsea Girl and Top Shop. Now, she shops mostly in Cheltenham, which is particularly rich in shoes. 'You can get good ones in Event, and Russell and Bromley.' Jilly's weakness for shoes was rewarded when she had to do a commercial for white wine. 'I got given masses of lovely shoes then, and they just go on.' Recently, particular favourites were the flowered, flat canvas pumps ubiquitous in the summer of 1985.

In Cheltenham Jilly also goes to Marks & Spencer for underclothes. She would consider spending a lot of money on such things a wild extravagance. 'If I was a millionare,' she says, 'I suppose I would enjoy expensive clothes because they can make you look wonderful. As it is, most of my successful things have been very cheap.'

Jilly is a natural observer of other people's clothes. 'I love people looking pretty and when they do I always say so, because I know how nice it is to be complimented. I think on the whole people dress to flatter the part of their anatomy they are proud of. I know my bosom is all right, and I have a small waist, so I make the most of them. After that, I deteriorate.' She thinks most of the young look 'stunning' these days. 'In fact,' she smiles, suddenly forgetting frilled dons' wives and wearers of strap-under-the-shoe trousers, 'I admire practically everybody's clothes. Paula Yates in particular, Candida Crewe, among the young. But being well dressed and having tremendous attraction often don't seem to be the same things at all. The women with most sex appeal never seem to be the best dressed.'

Jilly Cooper's friends have vivid memories of her appearance at parties: standing – perhaps by chance – against a light which brings to life a teasing silhouette beneath a clinging, plunging dress, she is admired and remembered for her stunning sex appeal. It seems fair, then, to suggest she might even cast herself in the category she defined: not the *best* dressed, but dressed in a way that is generally acknowledged as sexually appealing.

'Absolutely,' she agrees, delighted. 'I would definitely describe myself as a sexy dresser. That's what I dress for, that's what it's all about, to be sexy: to show that I can still be attractive, that I'm still keeping my hand in – that I haven't given up.'

The Honourable Pearl Lawson Johnston OBE, JP, DL

RACEHORSE BREEDER AND EX-HIGH SHERIFF OF BEDFORDSHIRE

'The High Sheriff's uniform gave me pleasure in that I felt I was looking the part. I liked it when people complimented me, though perhaps they were backhanded compliments. I think people commented because they were a little startled by the sight of me looking so smart.'

The Hon. Pearl Lawson Johnston was the first woman High Sheriff of Bedfordshire, from April 1985 until April this year. The appointment was something of a tradition in her family, her father, brother and nephew having all been High Sheriffs before her. The youngest of six children, (their grandfather was the inventor of Bovril) Pearl became a magistrate in her early twenties, in 1941. She served on the Bench for forty-three years, ending as Chairman of the North Bedfordshire Bench, as well as Chairman of the Magistrates Courts Committee. Her whole life has been one of active duty: again, at a very young age, she became county organizer of the WVS. By 1944 she was regional administrator for East Anglia with 67,000 members in her charge. Later, she worked in Japan during the Korean War, and for her services in the WVS was awarded the OBE. Her other appointments include County President of the St John's Ambulance, and Vice-President of the Bedfordshire Girl Guides. Always a keen rider, at one time she was joint master of the Oakley Foxhounds, and now runs her own stud farm, breeding from four mares. She looks after the horses entirely herself. A great music-lover and supporter of local music festivals, Pearl plays the organ in church and, in her few free moments, as a skilled needlewoman, she turns to the quiet pleasure of her tapestry.

'I have to admit I spend most of my time in trousers,' says Pearl. 'I have to get up and feed the animals, so it's the most practical. I don't consider myself a smart person, but I was brought up to be neat and tidy and I hope I still am. I was also brought up to wear the right thing at the right time so I always take the trouble, for instance, to find out whether I should be in long or short. My tastes haven't changed much, I just go gently along with the fashion: that is, I know what's going on, but don't take any notice if I don't like it. I don't like very way-out things, and there's so much that's *floppy*, now. That doesn't suit my figure. But the extraordinary thing is, I'll take something out of the cupboard that I haven't worn for a long time and I'll know at once whether it's too short or too long – I don't like

Right: Uniforms for women High Sheriffs are not officially provided. Left to her own devices Pearl Lawson Johnston used her St John' Ambulance suit and enlived it with silver buttons and lace cuffs. Her jabot, previously worn by other High Sheriffs in her family, was made from Bedfordshire lace. The hat was borrowed from the Mayor's Parlour and embellished with family Court feathers.

Overleaf: Pearl found this suit of man-made material in a shop in Bedford, the town she most frequently goes to for clothes. 'Not extremely smart,' she says, 'but frightfully useful for going out to lunch on Sunday.'

TRUTH WILL OUT
32
PLJ
MCMLXXIX
TRAVEL ISSUE

to be completely out when it comes to skirt lengths. I like shirtwaister-type dresses best. You could say my way of dressing is a little bit classical.

'I don't enormously worry about clothes, but I do find shopping worrying. Most people round here go to Milton Keynes, but I can do things quite well in Bedford. We've got of branch of Beales – I bought a very successful coat there, a light Norwegian business: heavy coats can be a drag. I also get a lot from Marks & Spencer. I never buy anything very expensive as I think it's unsatisfactory having things you can't dispose of, so I buy cheaper things that I can throw out without feeling guilty. I don't often go to London but if I happen to be there I might look round Peter Jones or Selfridges, and I might buy something if it took my eye. That's preferable to the feeling, "Oh gosh, I've *got* to find something for some special occasion", which is usually the case: my buying is mostly based on what I happen to be going to. As High Sheriff I had to attend a few official functions not in uniform, wearing just the ribbon of office, so I bought a smart summer dress which would do for things like garden parties and cocktail parties as well. As we never seem to have proper summers, my summer clothes go on from year to year: I'll only buy a new outfit if, say, there's a family wedding. I'm very keen on skirts and blouses and always enjoy buying extra ones. During my year as High Sheriff it was wonderful when I had to change in the evening because I thought up this costume of a long red skirt, one of my many white blouses, my lace jabot and badge of office, which is red and white, for decoration. But of course I can't go on wearing that now the year is over.

'I don't necessarily like something I've bought straight away,' goes on Pearl, 'but once I've got used to it I don't like parting with it. I tend to keep things for years. My favourite colour is pink – any pink but shocking pink. Funnily enough my racing colours, which I inherited, are shocking pink and white. I'll wear any colour, but prefer it to be bright. I have the inevitable black dress, and a black chiffon skirt which I wear with a tidy blouse for the evening. I don't mind synthetics particularly, but when it's very hot I like a real old-fashioned cotton dress. I like suits, but they're seasonable: a suit under a coat is very uncomfortable so you can't really wear one till May. I don't like polo necks or wide necks, but I don't mind frills. I always wear stockings, black with uniform, and quite often otherwise, or flesh-coloured. I don't like high heels and tend to wear a heavier type of shoe all the time. When I was High Sheriff I would drive to some function in flatties, then change into something smarter to get out of the car. Hats I only wear for weddings and funerals. I used to wear them for racing, but it's no longer necessary except for Ascot.'

Pearl goes racing regularly at Newmarket. 'In summer I wear a fairly thin dress and jacket,' she says, 'and in winter an overcoat. You see people at the Sales looking too anorakish for words,' she observes, 'but they're jolly smart by the time they get to the race course. A race meeting is a chance to show off nice clothes. Things worn at Ascot will come out again at York in August. I make a special effort on smart days. I usually wear my pearls, or a nice brooch. I'm very fond of jewellery and like to wear the odd bracelet or brooch in the evening. Amethysts are my favourite stone. When I go to a party that isn't very smart I always make quite sure that I'm wearing just the same jewellery that I would wear for a smarter occasion: I think it's very important never to dress up or to dress down.

'If I saw anyone who was very elegant I would recognize them as being so,' says Pearl, but the only person she could think of whose clothes she really admires is the Princess of Wales. 'When she came to Luton she wore a gorgeous magenta coat, it suited her so well. She has the height and the figure, of course, but she always seems to try. She never wears gloves, which is interesting. On the whole, I don't think Englishwomen do too badly. Those who didn't bother, or weren't able to, now can, due to the rise in standards. There are such lovely colours about, and so much to choose from. I've noticed some nice woollen two-pieces you can get now: the sort of people who wear them study what suits them. I think the young in their twenties are terribly sloppy, but those in their early thirties seem to be taking more pride in their clothes.

The Honourable Pearl Lawson Johnston

'I like dressing up,' Pearl goes on: 'Putting on the uniform was quite a performance – enjoyable, but I prefer dressing up for the evening. People do change for dinner round here. For something special, the family all together at Easter, for instance, it's always a black tie. I wear something like my long red skirt.'

Women appointed as High Sheriff are still obliged to put their own uniforms together. Pearl chose to wear her St John Ambulance black suit, to which she added good silver buttons and lace cuffs. Her jabot was made from Bedfordshire lace gleaned from her mother's lace collection. 'My mother was very keen on keeping the local industry going,' she says. 'It nearly died at one time, but luckily took off after the war. My brother was the one to have the lace assembled into a jabot when he was High Sheriff.' But now the term of office is over, the jabot has been returned to a glass case. The uniform hat belonged to the Lady Mayor: but as there was no Lady Mayor, and it just sat in the Mayor's Parlour in Bedford, it was loaned to her for the year. A black silk tricorn, Pearl embellished it with feathers. 'My sister-in-law's Court feathers tinted pink,' she explains, '*they* actually *did* a coming-out dance. The High Sheriff's uniform', she adds, 'gave me pleasure in that I felt I was looking the part.' And indeed she is one of that special breed of Englishwomen who look magnificent in uniform, evoking an aura of solemnity, dignity, duty and traditional, period elegance. 'I liked it when people complimented me, though perhaps they were backhanded compliments. I think people commented because they were a little startled by the sight of me looking so smart.'

MARIA AITKEN

ACTRESS AND DIRECTOR

'I didn't like red, but Jill Bennett gave me a red string sweater which made me realize I was wrong. As I get more like an old piece of Brie I've come to think red is a good idea.'

Maria Aitken's theatrical career began when she was at Oxford, where she was elected the first woman member of the Oxford University Dramatic Society. Since then she has worked ubiquitously – in repertory, on foreign tours, in the West End of London, and on television. Plays she has starred in include *Blithe Spirit*, *Bedroom Farce* and *Design for Living*. She began directing in 1982 – *The Happiest Days of Your Life* at the RSC, *The Seagull* at Greenwich, and William Douglas-Home's *After the Ball* among other works. In 1981 she founded her own production company, Dramatis Personae. Maria has appeared frequently on television as an actress but also, being a woman of many parts, as both a chat show host and a reporter in a documentary for the BBC about going up the Amazon. Her latest venture is writing: her book, *Adventuresses*, about life behind the scenes while making the documentary, has just been published. Maria Aitken lives in Kennington with the American architect Nathan Silver, and her son by a previous marriage. Cooking and keeping pigs she regards as pleasures when she has the time.

'Clothes,' says Maria, 'play a major part in the panic of life. Despite endless trips to the cleaners I never seem to have what I want sufficiently pristine, so I have to invent some new combination on the hoof. My entire wardrobe used to be taken up with stuff bought for television shows – not me, necessarily, but half price. I've always hated shopping and am notably bad at it. Three years ago I was a rotten, sweaty shopper bullied by salesgirls into making expensive mistakes. I used to go sales in grand shops – maximum hysteria-inducing, that. My only fashion tip is never go to a sale. Now, however, everything's changed, life's been transformed, by Gianfranco Ferre. If I'd met him at twenty-five it would have been a dream. As it was, my friend Roberto Devorik, who owns and runs a lot of clothes shops and was always hectoring me about my dressing, insisted I met Ferre. I went to a dinner that was given for him – in a dress by Bruce Oldfield, I remember. I thought Ferre marvellous and funny. He gave me a present from Milan – a white silky suit. I put it on and it was *exactly* right. That had never happened before. Now, I go to him all the time. Almost everything I have that's remotely smart is by him: beautifully made, simple things.

'I am vaguely aware of what's going on in fashion,' Maria goes on, 'I used to have a craving for fashion magazines. Busyness has put an end to all that, but I follow Susie Menkes who's very good and serious in *The Times*, and as Ferre dishes out stuff so regularly at least I'm in with *his* fashion thinking. I don't mind at all being unfashionable if I look nice. I usually cave in on the length of skirts. I have sloping Victorian shoulders so wear portable pads, even in my jerseys. I love them out of pure vanity – they put the body into proportion, though the trouble is I'm fated with lack of *talent* in shoulder pads: they work round to the wrong parts and I look like the Hunchback of Notre Dame. At heart, I'm a hessian skirt woman: I had to be *wrenched* away from my late sixties ethnic garments. I like to keep things for years, but I have got rid of things for Nathan.'

Unlike most Englishwomen in this book, Maria does not enjoy the process of dressing up. 'I have to

Right: Maria only dresses up with great reluctance: when she has to, a dramatic silk shirt, vast belt and simple skirt from Gianfranco Ferre are some consolation.

Overleaf: Very old trouser suit from Ferre, one of the first things she had from her favourite designer. Usually worn with a T-shirt, its status can be made to rise with a silk shirt for formal daytime occasions.

FONSECA
BIN 27
PORT

'Too much
good thing
can be

BRIAN JAMES

be *paid* to dress up,' she says. 'Bruce Oldfield did the clothes for my chat shows but I couldn't go through all that again. I could only dress for radio now. I didn't even dress up for my weddings. The first one, in a Manchester register office, I wore jeans with a pink coat that a friend hastily bought to cover them up. For the second I was in a Jean Muir dress, so restrained you couldn't call it dressing up. I'm often seriously under-dressed, never vice versa, and like it that way. On stage I just wear what the designers provide. Carl Toms is wonderful at designing stage clothes. One look at you, and he corrects the imperfections. He designs for the female body, as does Ferre. When I was acting in *Design for Living* I had constant pleasure from a beautiful white dress made by Yuki. And it was Yuki who gave me the bags I simply couldn't be without. They're huge, a cross between a briefcase and a bag, and I use them every day.'

The question as to what kind of dresser she considers herself to be caused Maria to think for a long time. 'A *comfortable* dresser,' she decided eventually, 'even if my comfortable clothes instil discomfort in others. But I find it hard to put myself into a category. I'm a chameleon, really. I dress first for work, second for me, though I would change something if Nathan objected. I wouldn't consult him in the first place, but if he didn't like something (and it's very demoralizing when he doesn't) I would change it. I'd probably brazen it out for the evening but wouldn't keep a high profile after that. I love appreciation. Nathan's pretty encouraging, but then he's American. I think they're better at it than the English.'

Maria has no structured approach to the buying of her clothes. 'I'm not really aware of buying anything at all, though of course I do,' she says. 'About once a year I realize how parlous I am sartorially, and I get down to fantasizing and making lists. I've drawers full of materials I've bought abroad, and have fantasies about finding a dressmaker. Apart from Ferre I go to Marks & Spencer for underwear, and at one point, when I had a satin pyjama fetish, I got a pair of satin pyjamas there. I don't have a shoe fetish: I love them but don't have as many as I'd like. I used to go to Katrina in South Molton Street, but that has sadly disappeared. Anello & Davide and Gamba are good for dyed satins, but I get most of my shoes in America now, about three pairs a year. I particularly like flat, very simple shapes. I bulk-buy tights – there's a new kind in C & A I like very much – they go up in a zig-zag pattern, very flattering. I sometimes press my nose against Lucienne Phillips' window and I get a few everyday things from Whistles. Ibiza, where my mother has a house, is the best place in the world for buying light-hearted summer gear. It's like St Tropez used to be years ago – two or three seasons ahead. Paula is the shop I go to there. They make their own silks. I get summer hats from them, too. I don't really wear hats, but I love them.'

The colour Maria would never wear is purple, and she used to think this also applied to red. 'I didn't like red,' she says, 'but Jill Bennett gave me a red string sweater which made me realize I was wrong. As I get more like an old piece of Brie I've come to think red is a good idea. I incline towards browns and creams and whites by day, anything but purple by night. Black stops me having to think, but it doesn't do me much of a service. Materials I like are silk and silk jersey, wool and silk, just wool, linen – but it's daft – and cotton is my favourite of all. My tastes have changed in some ways. I used to like ethnic things and I used to be keen on leather and suede. I could never wear a fur again because of the whole thought . . . more power to the ads. I hate bouclé wool, stuff with bumps, tapestry, and all the synthetics, though sometimes they're very clever: there's a silk jersey better than the real thing, but you can only get it in Japan. Really sophisticated synthetics can be okay, but obviously all that Crimplene material is impossible. I'm not keen on frills of any kind, or patterns, though I exclude classical things like dots. I rarely wear jewellery. One or two bits I like – I would have a penchant for earrings, but I don't have pierced ears. I'm hoping that evolution will happen. I love seeing jewellery on others.'

In considering the Englishwoman's clothes Maria admits to a 'Puritan streak about the whole event of dressing. While dazzled in Milan by the evidence of taste on every back,' she says, 'I find all that application rather decadent. Europeans make dressing a way of life, and I don't approve of that. I hark back to the vision of a gumbooted woman *doing* things. The English are novices when it comes to clothes – they seem to have neither the time nor the dedication. If they had more money and more time they could look wonderful, but they prefer to spend their lives bent double over a herbaceous border. I *don't* think they look good at balls, or at sporting occasions. They look completely frumpy, at night, compared with Europeans. Who do I *admire*? That's very hard. My mother is always wonderfully dressed and completely up to date. And Jill Bennett. She's very elegant without dedicating more than a fraction of her life to thinking about clothes. She leads an active life and always manages to look good: a lot of her things come from Jean Muir.'

Maria seems to be completely unaware of her own elegance. 'If you're over 5′8″ people always think you're elegant when you're really not,' she laughs. 'I'm nearly always dissatisfied with myself. Every year I make a resolution to do better, but I never have the time or the space. I would be happy for the rest of my life in floppy anythings. The secret of making dressing easy, I used to think, was to have a uniform, but I was never organized enough to get down to that. Actually, I'm always in a *sort* of uniform, rehearsing – track suits from The Gap in America, and boots – sluttish, day-to-day stuff. When I was directing a play in Chicago, one of the actors said to me: "You'd never know you were a success in England the way you dress."' It's hard to imagine which clothes in Maria's wardrobe would provoke such a remark. Perhaps it is the Puritan streak within her that blinds her to her own individual chic, the horror of thinking it a sign of decadence if you are preoccupied and satisfied with your own clothes. Her priorities, when it comes to her wardrobe, are very English, sympathetic, and much like those of the lady she admires – Jill Bennett. 'I think women should *try*,' she says, 'but not at the expense of anything proper.'

Previous page: Breeches and long sleeved shirt bought for a character Maria played some years ago in a television series. 'More for the character than for me,' she says, 'but half price.'

AGNES SALTER

NATIONAL CHAIRMAN OF THE WI

'I particularly love hats, which I know is a very anti the WI image these days, but still... People say it's difficult to make a speech in a hat, but I don't see why.'

Mrs Agnes Salter joined the Women's Institute in 1968. Having been on the national executive for seven years, and vice-chairman for two, in the summer of 1985, at the Albert Hall, she was elected national chairman. This means that five days a week she has to attend meetings in London and all over the country. But she still maintains her involvement with her own local WI and goes to group meetings, and she is still an *ex officio* market organizer. Married to a dairy farmer, Mrs Salter and her husband live near Bicester. These days she has little time to help on the farm. Her husband, on the other hand, helps out with the cooking, so she considers herself 'very spoilt'.

'For the meeting in the Albert Hall last summer I decided I'd like to get myself something really super,' says Mrs Salter. 'I went to Pickles, a new local shop in Bicester run by a friend. My husband came with me. I was after a dress and the very first one I saw was the right one. It was bright blue linen with a white trim, and had a hat to match. In the local shoe shop I found matching blue shoes, and I wore the outfit with white tights and white gloves. I love to have the right sort of bag, but never seem to. Mine always look tatty, but I do try. For the Albert Hall I found a white purse-type one. I wear gloves,' she adds, 'only for very special occasions – when shaking hands with royalty, for instance.

'I love clothes, and I always used to make most of my own. Now, of course, there's no time for that, so I have to buy. I find myself faintly thinking about them all the time as the necessity arises. If something big is coming up, I gear clothes to it. When I became chairman I didn't go out on a binge straight away: about three months after, I indulged in quite a concentrated buying spree. On the whole I wear pretty, casual clothes, although I've always had some smart ones. My wardrobe has always contained some basic mix and match things suitable for most occasions.

'I suppose you could call me a fairly classic dresser. I like simple, fairly classic things. I'm not really a frill person. I wear a lot of shirtwaisters. Trousers I wear at home, but never go out in them; dresses for dressy occasions. When it comes to buying I should say I'm an impulse buyer – if I don't find the thing straight away I know I'm not going to. I particularly love hats, which I know is very anti the WI image these days, but still.... People say it's difficult to make a speech in a hat, but I don't see why. I've got a pale aqua straw hat with a rose – a traditional summer hat, which is a great favourite, and I will probably keep wearing it for years. I'm inclined to buy things that won't date and I can wear often: favourite things go on for a very long time.

'I'm reasonably interested in what's in fashion, and will look at fashion magazines if the opportunity arises. I'll look at pictures and say: is that me or not? But I'm never swayed. I'll only buy what looks good on me. I'd never wear anything outrageous – I wouldn't be comfortable. I feel A-line type skirts suit me best. I've lost weight recently so now can wear a lot of skirts and blouses as I have a waist again. I have a very super Aquascutum tweed coat

Right: Mrs Salter likes dogstooth checks and stripes: this man-made dress combines both. It came from Leamington. 'I wear it for special meetings as it's quite dressy.'

Overleaf: For the AGM in the Albert Hall in 1985, when she was elected National Chairman of the WI, Agnes Salter chose her favourite colour, blue. She was lucky enough to find her idea of the ideal dress very easily in a local shop.

which I wear a great deal in winter. For evening I haven't anything very dressy as there isn't really any call for it. As my smart dresses are day dresses, if I do have to change in the evening, then I love long skirts and shirts. My favourite is an orangey-beigey acetate, patterned with sprays of orangey flowers, which I wear with a matching beige shirt.' On the whole, Mrs Salter prefers natural fabrics to man-made ones. 'I'd love to be able to afford them all the time,' she says, 'and I have quite a few real tweed things for winter and real cotton summer dresses. But some synthetics and mixtures are nice, and very practical.'

Mrs Salter feels her hair, which is a wonderful auburn, somewhat limits her choice of colours. 'I'm rather frightened of red, and tempted by pink but not quite brave enough to wear it. I wear a good deal of black and white, but blues and greens are my favourite colours. I used to like brown, but haven't worn much lately. I love coloured tights, and like to co-ordinate them with my shoes and clothes. My husband thinks I'm crazy when I wear green tights, but I love them. I dress for myself, though my husband is very straight about what I wear – he says it's either awful or jolly nice. I do make mistakes, but so far, touch wood, I've been quite successful buying clothes for this job.

'I enjoy dressing up and feeling on top of the world clothes-wise,' Mrs Salter goes on to admit. 'I like to feel I look good, it boosts confidence. People are kind enough to tell me when I look nice, and I like to be told.' With no hesitation she chooses the people whose clothes *she* admires most. 'The Princess of Wales. She always looks absolutely lovely. So does Princess Alexandra. And I can't help but admire Mrs Thatcher,' she adds with a smile. 'She's always well turned out – particularly good, I think, in black.'

A favourite pure wool skirt, blue jersey and shirt: the sort of comfortable clothes Mrs Salter wears at numerous meetings of the WI she attends all round the country.

Roses

The Viscountess Rothermere

One of the Last Romantics

'People sometimes think I'm flamboyant, but really my clothes are just very simple and comfortable.'

Lady Rothermere and her husband, who is Chairman of Associated Newspapers, have four children. When in England, Lady Rothermere divides her time between her magnificent flat in Eaton Square where, as one of London's most generous hostesses, she gives legendary parties, and her house in the country. She also travels frequently between her house in California, her flat in New York, and her house in the South of France – which once belonged to Greta Garbo. She is involved in various businesses and works for several charities, and somehow maintains the extraordinary energy required for this hectic life. But, as an active mother, Lady Rothermere's chief pleasure is to be in the country, or in Jamaica, with her children.

'Ever since I was young I've had an image in my mind of how I wanted to look,' she says. 'As a teenager I designed my own clothes – *fluid*, plain things with long sleeves softened with frills. I still sometimes give people design ideas. I've never been confined by the latest look, although I do vaguely keep in touch with fashion and look at things in magazines. I'm definitely a romantic, but *individual*, dresser. I love dressing up, but I also love country clothes. Yves St Laurent makes them for me – tweed skirts and jackets and lovely capes: sometimes I have the same shape in two or three different materials and colours. And I've got a lovely country suit from my husband's tailor, Addison & Shepherd, wonderfully warm in cold weather. I wear it with a shooting hat. I don't much like ordinary city day clothes, and just wear very simple things, mostly skirts and shirts, usually white silk. I wear them with those Pringle cashmere stoles, which I love and have in several different colours.

'I practically never go shopping because I hate it, but I do go and see collections – Gina Fratini, Zandra Rhodes, Jean Muir. I've got clothes from all of them I've had for ages – in fact I hardly ever throw things away but wear them for years if I like them. I have lucky clothes, dresses and shirts I usually have a good time in. Others are unlucky – I never know why. Often I feel I've made a mistake, usually about the colour. Burgundy is my favourite colour of all, but I also love some pinks, red and black. I *don't* like purple, and certain nasty pinks. Tights I usually have in black or beige or blue. I like Italian shoes best – simple, comfortable and flat. I get a lot from Magli. I have to have large handbags that I can stuff full of everything – Gucci, usually. But I'm not a hat person. I never wear them except for openings, and then they're large and romantic from John Boyd or Freddie Fox. I have to admit I do love rich, glamorous fur coats, particularly sable. I've quite a few fur coats. I don't wear much jewellery though I've a few good pieces, mostly old. I love rubies, or just plain diamonds. I also have some modern chains from Bulgari and Boucheron.

'The sort of dressing I don't like is that uniform look that the French, especially, go in for – those tailored women who look elegant but are all the same. I admire Lady Hambledon's clothes – but of course she's Italian. Englishwomen's dressing on the whole is... *unremarkable.*' But she did manage to think of just one woman whose clothes she admires: Lady Rupert Nevill, who, she says, is 'always beautifully dressed'.

Although Lady Rothermere's collection of clothes is large and impressive – she keeps basic things in all her various wardrobes – she does not

To Lady Rothermere dressing up at night is a great pleasure, and indeed she is renowned for her beautiful evening dresses. This one is wild silk in her favourite colour, burgundy, from one of her favourite British designers, Gina Fratini.

THE VISCOUNTESS ROTHERMERE

spend a great deal of time thinking about them. And although she hates shopping in general, there are a few shops she enjoys, such as Chervet, in Paris. 'A divine shop,' she says. 'They have beautiful, mannish coats for women – I love masculine clothes on feminine women – and marvellous silk ties which you wear in your hair or round your neck.' In New York Oscar de la Renta and Bill Blass are her favourite designers.

In common with many Englishwomen, Lady Rothermere's greatest pleasure, when it comes to clothes, is to dress up for the evening: she is renowed for her exquisite, often elaborate, ball dresses, many of them by Zandra Rhodes, Gina Fratini and Yves St Laurent. She particularly likes three-quarter length evening dresses and nineteen-thirties shapes. Her favourite materials for evening are romantic – 'chiffon, lace, paper taffeta, in pale pinks and creams, boudoir colours, and also burgundy. I dress for myself,' Lady Rothermere adds, 'and I do take trouble. I like to feel I've got the clothes I'm going to look and feel good in – to be happy with what you're wearing gives you a terrific amount of confidence. Hair, I think, is terribly important: you should always have it done because that's a great morale-booster. But in England there seems to be little *demand* to make an effort, whereas in Paris you simply have to. Over here women seem to think they can get away with being under-dressed, and not trying very hard. I myself love seeing other women look nice, and always compliment them: in fact, Englishwomen seem to earn more compliments from each other than they do from men. My own idea is always to be elegant and generally well groomed. People sometimes think I'm flamboyant, but really my clothes are just very simple, and, most important, comfortable.'

'Ever since I was young I've had an image of how I wanted to look,' says Lady Rothermere, and thinks an apt adjective to describe herself might be 'luscious'. Luscious dress from another of her favourite designers, Zandra Rhodes.

ELIZABETH JANE HOWARD

NOVELIST

'It's either evening things or culottes. There's nothing much in between. I'm really bad at sudden lunch with a colonel – no good at clothes for the halfway thing.'

Elizabeth Jane Howard, the novelist, has been married (she has a daughter and four grandchildren) but now lives on her own in Camden Town. She is working on a trilogy of novels, and a cookery book, which she wrote with Fay Maschler, will be published shortly. Her life is a disciplined one, writing five to six hours a day, and she rarely leaves London except for the odd weekend. An excellent cook herself, she tries to give fairly regular dinner parties for anything up to sixteen, all of which she prepares without help. In any spare time, she makes intricate shell frames for pictures and looking-glasses; she also does patchwork and tapestry, and knits beautiful jerseys for her grandchildren.

'The trouble with me,' says Elizabeth Jane Howard, 'is that I love clothes but I'm having to come to terms with the fact that I'm no longer sylph-like. I've no waist any more – I'm altogether a larger, different shape. So I'm in the process of changing my style, and in that process I make a lot of mistakes. I get about fifty per cent of my things made by two dressmakers – I used to have a Polish one who I went to for thirty years. It's almost useless for me trying to buy things off the peg in London, though it's fine in Paris or New York. I do sometimes go to sales at Chic in Hampstead, or get something from Jean Muir. But relatively cheap clothes for my size don't exist, which is very depressing.

'What I'm doing now is feeling about for what I find comfortable and suits my life. I was never very fashion-conscious, but a romantic dresser, which I can't be any more. I find myself still wanting clothes that no longer suit me. For instance, I used to wear a lot of straight skirts with shirts: I can't do that any more. In trying to sort things out I've become a very erratic buyer – an anorexic person about clothes, you could say: I don't think about them for weeks, then I can't stop buying. I've no organized plan, I'm rotten at it really. I'll go out and buy a sweater I don't need just because I'm feeling gloomy. I'll go to Allans' sale thinking I must buy some grey flannel, and come back with something quite different. I do look at fashion magazines in the hairdresser, but I don't care what people are wearing this year, and anyway there are no designs for people like me.'

Culottes, Elizabeth Jane Howard has found, are one answer to someone her size, comfortable and practical. She has several pairs in various materials and wears them much of the time with large, beautiful jumpers, many of which she knits herself. She wears them quite long, as she does all her skirts, remembering Diaghilev's observation that 'women's knees are the least attractive part of a woman's anatomy'. She claims she is 'really into' long jackets and wants more, and loves coats, though she only has two, and one of those, from Jean Muir, is twelve

Right: 'My wardrobe's full of mistakes,' says Elizabeth Jane Howard, but this isn't one of them: poppy painted black velvet from Allans made into a jacket copied from an Italian design, with silk-satin T-shirt and velvet culottes, 'all only two years old' – young by Jane's standards.

Overleaf: Jane works from ten till two every day: this is customary working gear. 'The sort of thing in which I can go straight from the typewriter to gardening or cooking.'

The Good Gardener
REMSTAR 101

Elizabeth Jane Howard

years old. She also has a particular penchant for mackintoshes and is enthusiastic about a fairly recent discovery: it's shiny black stuff with a tartan pattern and a warm cotton lining. 'I feel like the head of the Mafia, or Women's Lib in it,' she says. 'I like it so much I've bought another one, just the same stuff, in short.'

There are few day dresses in Elizabeth Jane's wardrobe. 'I find them difficult,' she explains, 'chiefly I hate being hot, so like to wear layers that I can discard. I only have one good day dress, I think. What I really love is dressing up for the evening – but there again I'm having to come to terms with the fact that I don't need so many evening things these days as there aren't so many parties. I must say I miss all those lovely long skirts and bare tops – though I still wear lowish tops as my neck hasn't gone yet. Now, really it's either evening things or culottes. There's nothing much in between. I'm really bad at sudden lunch with a colonel – no good at clothes for the halfway thing.'

In the process of Elizabeth Jane's sartorial metamorphosis, even her taste in colours has, surprisingly, changed. 'I used to be deeply devoted to olive greens,' she says, 'but now I don't want them at all any more. Now, I like acid greens and pinks, which I would never have worn before. Also, a lot of black and white. The materials I love have always been velvet and silk and cotton jersey, things that don't look awful in five minutes. I had a blue linen suit from Jasper Conran that looked dreadful almost as soon as I put it on. I *don't* like man-made fibres, but the trouble is that the materials I do like are so expensive. I go to sales at Allans' and Liberty's and Jason's in Bond Street, and sometimes I'm lucky. But what strikes me is what you *can't* find. I'm not *ravished* by materials any more.' Despite her despair caused by declining materials, Elizabeth Jane Howard does, in fact, have a talented eye for spotting beautiful stuff, and has some memorable velvets, for instance, made into evening jackets and skirts. They include a black velvet jacket printed with poppies, and a stamped velvet trouser suit with eighteenth-century buttons – 'the kind of thing that will last for the rest of my life'.

'On the whole,' she says, 'I'm spurred by an occasion to buy something new.' She is lucky enough to have Haachi as a neighbour, and for a summer party given by Dru and Jack Heinz some years ago, he made her an exquisite long white satin skirt which she wore with a black beaded top. She also bought from him a black, self-patterned silk Grecian-looking evening dress. 'Those will go on,' she says, 'things I love I keep for ages. I've still got a marvellous Wedgwood blue Victor Stiebel ball dress, twenty years old: putting it on was like getting into Fort Knox. I had some lovely parties in it, and now I keep it for lending to grand-daughters.'

Chic in Hampstead is one of the few shops in which Elizabeth Jane Howard feels happy, and where she is often successful at buying. 'I managed to find a very good shaped skirt there – something that's almost impossible if you're quite large. Of course, it was expensive, but it was worth it. The nice thing about Chic is that the sales ladies know what you want and are very helpful without pushing you into anything you don't want.' Buying shoes, though, Elizabeth Jane Howard finds a very depressing matter indeed. 'I mind frightfully about shoes – comfort is vital. On the whole I buy most of mine in New York. The trouble is my feet aren't pointed, and most English shoes are pointed and low cut so your toes show – hideous. I don't like high heels and I don't like boots, and I can't find Idlers any more. So I dread going out and looking for shoes: they're so expensive that to make a mistake is awful. It's almost impossible, too, to find tights over here that fit anyone with long legs. The English are hopeless at long legs. So I have to wear stockings, or get my tights in New York, where they come in wonderful colours and *fit*. I like very fine stockings in beautiful

From Chic of Hampstead by Sheridan Barnett: Jane describes this dress with the same precision as she describes clothes in her novels. 'Full and rather dignified, slightly stiff silk, not pink, not cream, but that colour that faded roses sometimes are.'

colours. I always have dozens of pairs and do think about them a bit. As for hats, I had a lovely red fedora from Herbert Johnson, which was stolen on a bus: now, I only have one I ever wear, of real fur.'

Disenchanted by her own clothes though Elizabeth Jane Howard may be, when it comes to jewellery quite a different light comes into her eye. She is *famous* for her jewellery. Her collection includes some eighteenth-century paste, but is mostly ancient gold. 'I've got various Roman chains including a very beautiful one indeed, two thousand years old, and my oldest earrings date from 4000 BC. When I first started buying ancient jewellery it was very cheap. Now, it's much harder to find anything affordable. I don't like junk jewellery and never wear it, though I like it on others. It's interesting, those who notice my jewellery and those who don't. . . .'

Elizabeth Jane Howard's view of the Englishwoman's clothes in general is that they are *rotten*. 'But not rotten for the same reasons as me,' she says. 'I think, mind and sometimes enjoy, although I don't always get it right. No: on the whole, Englishwomen expect to go about looking inconspicuous, like hen birds. "I think this is all right, I think this will do," they say to themselves. Molly Norwich's daughter once said to me, "You should wear your best clothes every day," and I was struck by that. I suppose I've always dressed for myself, though when I was young I dressed to attract men. Kingsley, when I was married to him, was never very interested, although I always wanted him to like what I was wearing. I do like to be complimented: I always tell other women when I think they're looking good. I think you should – though there aren't that many occasions to compliment Englishwomen. I think, in general, they should try harder. They *can* look wonderful, at grand balls. But it's those people who think if their shoes match their bag they are smart who get it wrong: that's absolutely not what smartness is about. Children are better dressed now, I think, but the improvement doesn't extend to their mothers.'

The exceptions Elizabeth Jane Howard can think of to this general rule are Tamasin Day-Lewis and Anne Scott-James. 'Tamasin has a real panache about clothes - it's the way she wears things with an air that makes you feel that it's just right. And Anne has a great air, great style, too.'

The whole perspective of clothes in Elizabeth Jane Howard's life has changed over the years. 'They're much more important now than when I was young,' she says. 'I was very poor then, but it was easier to dress because I had a good figure. But nowadays,' she reflects a touch ruefully, 'I have to give the whole business much more thought. At the moment, I'm rather an eccentric dresser, it's all rather a hit-and-miss business. I make masses of mistakes, some of my clothes are really crumby. But I'm going to practise. I know I can't hope to look wonderful all of the time, but I'd like to just some of the time. I'm going to get it right. Because, then,' and she smiles at the happy thought, 'thank goodness, I won't have to think about it all so much.'

VIRGINIA HOLGATE

SPORTSWOMAN OF THE YEAR IN 1985

'Jewellery I absolutely adore. *I've always wanted a pretty brooch – a frog or a bee or a hippo or a pheasant or an elephant. I'd be quite happy with any animal.'*

There can be few English uniforms so flattering and romantic as side saddle clothes for hunting and the trim tailored jacket and jodhpurs for pure dressage, which Virginia Holgate considers 'absolutely charming'. All her riding things are provided by Harry Hall.

Virginia Holgate is surely the heroine of all those English children who spend their childhoods on ponies and dream of riding for England when they grow up. She began her equestrian career at a very early age, and at eighteen won the Junior European Three Day Event Championship. Soon after her twenty-first birthday she had an almost fatal fall, breaking her arm in twenty-three places. After the arm had miraculously healed, she returned to a career in three-day eventing that has proved remarkable by any standards. At the Olympics in Los Angeles in 1984 she won a team silver medal and a bronze individual medal. In 1985 she was winner of the European Championships, the Burghley Horse Trials for the third time, and also the Badminton Three Day Event. Her two advanced horses are Priceless and Nightcap, both sired by the famous stallion Ben Faerie. To add further to the accolades, last year Virginia was voted Sportswoman of the Year by the Sports Writers' Association. At the end of the year she married Mr Hamish Leng, a financial consultant. Sponsored by the British National Life Assurance and clothed by Harry Hall, Virginia spends much of her non-riding time giving lectures and demonstrations all round the country. She and her husband have a flat in London and a cottage near Badminton, adjacent to her mother's house and her horses' magnificent stables.

'I can't afford all the clothes I'd like,' says Virginia, 'but they're one of my favourite occupations. Both my mother and I are very fashion conscious and have our own personal thoughts about being well dressed. I love simple, chic clothes – suits and trouser suits – I very rarely wear a dress in the day.'

'Well, there aren't many dresses we like *made*,' observes her mother. 'My mother,' smiles Virginia, 'battles away saying don't buy rubbish, so I try to buy something really good and hope I won't have to buy another of whatever it is for ten years. That *is* beginning to work. I definitely make mistakes, but not horrendous ones – it's usually the colour of shirts or sweaters. I like all the evening dresses I've ever bought and still wear them. I'm into very simple things at the moment, nothing too colourful. I like greys, yellows, rich dark blue – my favourite is that purply colour, lilac – a *good* lilac. And I like a *good* peach, a sort of azalea. I keep off strong colours like shocking pink, and I don't go for green or red, and very rarely wear black.

'My tastes have changed. I started off liking simplicity, but at twenty-two had turned to more fussy, flowery things. Now I'm back to the original style, trousers and a smart jacket. I'm a classical

Virginia's going away dress last year was a very 20s number, made of angora and 'surprising' to her husband. She keeps things she likes a long time: this dress was intended for many an occasion beyond the wedding day.

dresser, I would say. I like everything really waisted or dead straight. I look dreadful in V-necks or round necks: I must have very high or very low necks, and I love strapless evening dresses. I do take a bit of notice of what's currently fashionable but adapt it to myself. It's no good my wearing very wide shoulders and a long jacket – I look like a dwarf. But I love shoulder pads so long as they're not too big. I don't like sequins or sparkly things. I try not to buy synthetics but as always it depends on money. If I could afford it I'd like a lot of cashmere, real satins and silks. I love pure cotton for shirts, it feels gorgeous and looks so nice. Ideally, I'd love to buy new shoes with each outfit, but I can't. I've some simple courts in patent leather and find I can get away with them most of the time. There's not a bad choice at Russell & Bromley, though it's hard to find exactly what you want. Colours out of season you can't find for love nor money. I think nice boots are essential, though I haven't got any: I'd like plain black ones with a brown top. I have just one pair of plain silver court evening shoes, and that's it: I make do with what I've got, and slowly collect. I do look after my things very carefully – shoes in trees, handbags put away in plastic bags. I've only one overcoat in a cashmere and wool mixture. What I long for is a real camel coat in beige-cum-cream. I'll have to wait for that. . . .'

Owing to her peripatetic and extraordinarily busy life, Virginia has just four hours a week left for her entire shopping. 'If I have a lucky day,' she says, 'I really enjoy it. But I'm so pressed for time that usually I find it very frustrating. I *never* have time to browse, except off season, from November to January. I try always to keep a look-out and if I suddenly see something spectacular I buy it, even if there isn't an immediate occasion for it. The two cities I mostly shop in are Bristol and Bath. I go to Tiziano in Clifton and Image in Bath: they're both designer shops, selling people like Roland Klein, Gina Fratini and Cacharel: nothing dramatic or fancy – simple, well-made stuff. I know both shops well and they're very helpful – they'll ring up and say if they think they have something I might like, and I ring to see if they have something particular. They'll say yes or no, which saves a lot of time and wasted trips. In London I don't go anywhere specific. I find Harrods mesmerizing and don't know where to begin to look. I like Harvey Nichols and Simpsons, Marks & Spencer for polo-necked jerseys and underwear, and C & A for very good skiing kit, not too expensive.

Virginia's riding clothes are supplied by Harry Hall. For everyday, when not in jeans, she lives in their riding breeches, which come in blue, green and black corduroy. Beautifully cut and thigh-clinging – rather than the unflattering shape of traditional jodhpurs – they are surely the solution for all women desperate to find good trousers, even if they have no intention of mounting a horse. 'I find trousers a nightmare,' says Virginia, 'having a small waist and hips, then rather going out.' Harry Hall's cord breeches are the flattering answer – she has them in several colours. The firm also provides her stocks, shirts, cream show breeches and riding hats – 'though we now have to wear crash helmets,' she says, 'which is sensible but rather sad. For pure dressage I think the clothes look absolutely charming – cream breeches, with tail-coat and top hat in black, and I wear a false bun. I think women look wonderful hunting in side-saddle clothes. I shall enjoy hunting when I retire. I enjoy riding kit – it's very comfortable.'

Many of Virginia's everyday clothes come from a small firm called Good Going, which sets up a booth at horse trials and shows. 'Frankly, we do a lot of our shopping with them,' she says, 'it's so convenient. I get their jeans and T-shirts and they do sweaters in lovely colours, all different designs – words all over them if you want, they can do anything. You order whatever you like and it arrives in about a month.'

When she has time off from riding, Virginia likes going off to London for a lunch or a meeting. 'I *love* dressing up for that sort of thing,' she says. 'I'll usually wear a suit or trouser suit and silk shirt. I haven't *any* winter day dresses, though I do have a blue silk summer one with a high neck and buttons that I adore. I wore it on the Terry Wogan show. I'm currently looking for a cocktail dress: that's *very* difficult to find. I also love dressing up for the evening, though sadly that happens more in London than the country. For a black tie dinner, say, my first reaction would be to wear my Gina Fratini cream wool breeches lined with silk and a matching silk shirt. I often wear smart trousers and a lovely shirt for dinner. I think they look super, with short boots or patent shoes and nice earrings and a brooch. Jewellery I absolutely *adore*: if I had money I'd go bananas. I love earrings and have always wanted a really pretty brooch – I'd love a frog or a bee or a hippo or a pheasant or an elephant: I'd be quite happy with any animal. I'm hoping my husband, in about ten years' time . . . I wore Hamish's family tiara for my wedding, and have a beautiful diamond engagement ring. I love the pearls my father left me for my twenty-first, and I love sapphires and rubies, though I'm no great fan of emeralds. I do wear junk earrings, Butler & Wilson and Monty Don.'

It was from Gina Fratini Virginia bought her wedding dress for her marriage in December 1985. 'I walked into the room and there it was,' she said, 'heavy ivory silk, puffed sleeves trimmed with lace, high neck lined with lace, a lace-trimmed bustle, and a bow at the back: very simple, rather Elizabethan. Gina is one of my favourite designers.'

At race meetings at Ascot and Cheltenham, Virginia thinks, you do see some well-dressed women, and you *can* walk down a street and come upon occasional elegance. 'But that's not often,' she says. 'In general I find Englishwomen's dressing abysmal. Women of thirty upwards dress a bit better than they used to, though they couldn't begin to hold a candle to the Swiss, French, Spanish and Italians. I do think it's essential to grow old elegantly. The English probably do care, but they just don't have enough time. The young I can't make out at all – perhaps the Princess of Wales will have some influence. She's got *incredibly* good taste, I think. If I'd had her figure and her opportunities, I'd be a good dresser. . . . I much admire her. She never overdoes it, is always elegant, classical, chic and shows great imagination. I also admire Lady Joanna Norman who used to have a boutique in Walton Street and designs for herself and her daughter. She's thin as a rake, and stunning. If she was still in business, I'd go to her. I also love the way Virginia, Lady Leng, my mother-in-law, dresses, and of course my own mother. They both stood out a mile at the wedding. Mummy helps me a lot in choosing things. I'd never buy anything serious without her.'

Who does Virginia dress for? 'Well,' she says, 'I think men are right when they say women dress for themselves and for other women. Hamish does notice how I look and sometimes he says, "Gosh, you look nice", but he isn't a bore, bleating on about my looking marvellous when he hasn't a clue. I must say he went berserk about my going-away outfit – an azalea angora two-piece with a matching headband. I think it quite surprised him because, really, he thinks I'm a bit of an urchin.'

Previous page: Virginia likes dressing up at night. This is her favourite evening dress – six years old from Lucy's of Knightsbridge.

THE LADY ALEXANDRA DACRE

WIFE OF THE MASTER OF PETERHOUSE, CAMBRIDGE

'I only wear a kilt and a tweed jacket when I go to the Highlands – you're not meant to wear them in the Borders, though I did put on my kilt when my cousin General Alexander Haig came to lunch with my brother.'

Lady Alexandra Dacre was undoubtedly the most elegant academic's wife in Oxford, where she lived for twenty-six years. But in 1980 her husband, the former Professor Hugh Trevor-Roper, was made Master of Peterhouse, Cambridge. There, for the last five and a half years, she has been equally acclaimed the best-dressed wife of a head of college. Always a great lover of music, Lady Alexandra founded the Music Therapy charity in 1970, but resigned from the chairmanship two years ago: she has remained a governor. She is the Patron of the Cambridge University Opera Society, and goes frequently to the opera in London. A skilled designer of gardens, she has done much to develop the one at the Master's Lodge, and over the years has created an enchanting garden at the Dacres' house in the Borders.

Lady Alexandra is an instinctive dresser who, every day rather than just on important occasions, gives thought to her clothes. So even at the most unexpected times, you can never catch her out: she maintains a consistent elegance, has a marvellous eye for colour, and imbues even inexpensive clothes with an aristocratic air.

'When we were children, my sister Doria couldn't bear clothes,' she says, 'nor could my mother, who dressed us in appalling things, much too big for us, which really upset me. But then we went to live with an aunt who dressed us beautifully: it was terrible when we returned to my mother who continued to dress us in things quite unsuitable for small girls.'

'I began to choose things for myself when I came out at seventeen. I particularly remember a green chiffon dress with a water lily at the waist, and a navy-blue taffeta day dress which I liked very much. Favourite things I remember vividly for years. There was a white satin evening dress with amber embroidery and a bustle made by Jacques Fath when I lived in Paris. And I still have some things from Annacat – everything she had suited me. I think I really began to enjoy buying when I lived in Paris.' (At the time she was married to the Naval Attaché at the British Embassy.) After the war I was dressed by Jacques Fath: I was tall and thin which is what he liked. Once he designed a dress actually on me, pinning it as I stood there – years later I gave it to the Victoria and Albert Museum. My *vendeuse* at Jacques Fath was Raoul Dufy's sister. I've always loved hats and bought marvellous ones in Paris. I found some beautiful ones at a place called Suzy. I remember a shiny green straw – same green as the chiffon dress – with a crown of lilies-of-the-valley and a pretty veil. I love veils on hats and wear a lot of them. Another hat from that time, white silk with a flowered brim and a black velvet crown, I've only just given away, again to the V and A.

Today, Lady Alexandra's enjoyment in clothes still thrives, but she finds shopping more difficult. 'It's so confusing, shops filled with things quite out of season. I buy much less now but I do plan and about twice a year I think about adding to my wardrobe. Then if I'm going to some special occasion, I really enjoy thinking about what I'm going to wear: perhaps if it's a lovely day I just go off and find something new. I used to go a lot to the Dior

Alone in elegance among academics' wives: Lady Alexandra in an aubergine wool crêpe Jaeger suit with printed Viyella shirt and separate tie. Variations on purple are her favourite colours.

boutique in London, but that sadly closed. You can find very little in Cambridge, but I go to Jaeger, and to Next for everyday things, and Marks & Spencer for jerseys and underclothes. If I want something special I have to go to London, often to Sarah Spencer in Beauchamp Place. I bought a wonderful scarlet cloak there. I particularly love cloaks and have several of them. They're practical, and never date.

'I do keep in touch with what is fashionable, but it doesn't much influence me. I simply wear what suits me, and I'm quite happy to wear old things if I like them. I've still got clothes from Lanvin that are twenty years old. I like good materials: real wool, velvet, silk. I never wear synthetics if I can help it. I enjoy looking for materials and getting things made, and would get many more things made if I could find a dressmaker nearby. One of my recent favourite dresses is of beautiful scarlet and white silk that I found in Liberty's: I got the idea for its shape from a magazine, then did a drawing and had it made by a dressmaker in Oxford. Another is a long dress made from lime green and white Thai silk that Hugh brought me from America. That one I designed myself.'

Previous page: Chosen for her daughter's wedding at Christ Church Cathedral, Oxford, in 1977, and still often worn: printed crêpe de chine dress, purple wool cape bound with mauve, hat made especially to match by Freddie Fox, with spotted veil.

Right: Seemingly uncrushable Liberty's silk found by Lady Alexandra and made by a dressmaker from a design in a magazine. 'Good both for garden parties and dinner parties,' she says. Scarlet face-cloth cape from Sarah Spencer.

The Englishwoman's Wardrobe

Looking for clothes, Lady Alexandra is one of those people who makes up her mind immediately. 'I know at once what I want. I never dither. My favourite colours are mauve and various shades of purple, green and red. High necks suit me, frilled so long as they're not too overdone. I think I'm rather a romantic dresser. I like tight waists and longish skirts. I can't wear very masculine clothes. I only wear a kilt and a tweed jacket when I go to the Highlands – you're not meant to wear them in the Borders, though I did put on my kilt when my cousin General Alexander Haig came to lunch with my brother. I thought it would be expected on that occasion. In the country I wear corduroy trousers in the winter, and lots of tweed skirts with pretty shirts and cardigans. I definitely like winter clothes better than summer clothes. In London I'm inclined to wear something dark, navy blue perhaps, though never in winter. Gloves I wear on special occasions, suede or leather or white cotton. Hats on special occasions, too, like when Hugh took his seat in the House of Lords, or when my daughter was married. I get them mostly from Freddie Fox in Bond Street, now. He usually makes one of his shapes in a special colour for me.'

It was on the occasion of her daughter's wedding at Christ Church Cathedral, in 1977, that Lady Alexandra appeared in some of the first spotted tights to be seen, mauve to match her dress and cape. She often wears various shades of purple tights, but also other colours to match whatever she is wearing, and flesh-coloured ones to go with light things in summer. Shoes are her difficulty, having extraordinarily narrow feet. 'I used to go to Charles Jourdan, now to Ferragamo, but none of them last very long considering how terribly expensive they are now. As are bags. Mine have to last a long time so I always choose classic shapes.'

Lady Alexandra loves antique jewellery although she does not wear much in the daytime. Amethysts are her favourite stone. She has a collection of Victorian rings and brooches which, with her knack of finding pretty things that are not too expensive, she has gathered in the last twenty years. For special occasions she sometimes wears a favourite Cartier watch that belonged to an aunt. 'And I have a few lovely things from my godmother, Queen Alexandra,' she says, 'including a pendant made with a diamond double A (Queen Alexandra always used the cipher as Alexandra begins and ends with 'a') engraved on an amethyst.'

Thinking about Englishwomen's clothes in general, Lady Alexandra's view is that they're 'frightfully dull'. 'Either they're a mess, or trying to be on the safe side. I like French clothes much better,' she says, 'and the Americans in New York are *soignée*. I rather like some of the *Dynasty* clothes, although they're very theatrical – they wouldn't suit many English people.'

Whose clothes does she admire, if any? 'Madame Giscard d'Estaing's and Lady Berlin's – but then they're not English.' A long pause while she tries to think. 'Lady Gladwyn,' she comes up with at last, 'who used to dress at Fath. And the Duchess of Kent and Princess Alexandra.'

Lord Dacre sometimes gives his wife clothes. 'But of course,' says Lady Alexandra, 'I always choose them myself. I dress for myself, on the whole. Hugh, I know, likes me to be well dressed but he never actually says anything. I'm often prompted to say "Have you seen this before?" In Cambridge, everyone's so dowdy I expect I often look over-dressed. But Hugh, when I ask him, says he likes me to dress as I do and wear hats.' Modesty forbids Lady Alexandra to say how frequently she is complimented on her clothes. She will merely admit that when people say they like what she is wearing she can't help feeling 'terribly pleased'.

Miranda Brett

EX-OXFORD UNDERGRADUATE, NOW WORKING IN THE CITY

'I have a sort of dream . . . I'd like to go to Harrods for an appointment with Mr X who'd found clothes for me that would be right!'

Nostalgia for undergraduate dressing persists: at home it's still black leggings, Joseph Tricot jersey and gym shoes from Woolworth's.

Miranda Brett was something of a Zuleika Dobson during her three years at St Anne's College, Oxford. She came down in 1985 with a degree in physics. Three months later the carefree days ended with the kind of shock she could never have anticipated: the disciplined life of work in Hambros Bank. Now a reluctant commuter by underground to the City, her long working hours do not mean she has abandoned parties at night, and she often visits old friends in Oxford at weekends. She has no idea what the future of her working life will be.

'I've always thought a hell of a lot about clothes,' says Miranda, 'chiefly because I'm never sure what I want and I'm never satisfied with what I've got. I mean, I'm in tears about twice a week. I find myself spending far too much money on rubbishy things instead of a few good ones. I buy things I like that don't suit me. I'll go shopping for a whole day and

BARBICAN CENTRE
tokyo
march 83

come back with nothing – there's very little around at the moment I like. And then maybe I'll reflect a bit on what I've seen and go back and get something I wasn't sure about in the first place and *that'll* be wrong....

'I didn't used to be like that. I used to *know* what I wanted, at Oxford and before. I would save and buy, no problems. Though I have to admit in Oxford I would go and buy something tacky just to cheer myself up and then never wear it. I like to think that at Oxford I dressed as I liked, but in fact there was a kind of uniform we all conformed to.... We were a pretty scruffy bunch and all liked the feeling of being scruffy. Most of the time I wore black leggings, huge jerseys, ankle socks and ordinary black gym shoes from Woolworth's. In summer, T-shirts with the leggings. We'd *never* dress up unless specially asked to. I'd make an effort make-up-wise, but never wanted to *look* as if I'd made an effort. My first year I wore a lot of bright colours, pink and red. The second year I became more trendy and bought a lot of black and grey, which I haven't been able to throw away. I still like black, and electric blue and pink. Never orange – though in fact I have just got an orange jersey, which I wear a lot for some reason. I'd never buy that really nasty hard pink and the equivalent green.

'Home clothes don't change. It's still the leggings and gym shoes, and a lot of very mini skirts which I've been wearing since I was fifteen. Actually I do wear them to the knee now, but don't feel comfortable. Dresses are the great problem. I love the idea of them in theory, but I can't find one. A day dress, I mean. I've been searching for three years and I haven't even the picture in mind of the kind of thing I'd like. I do love dressing up, when there's a chance. For the odd black-tie party, for the past year I've worn a very mini black velvet dress from Photo in the King's Road – very tarty, but I'm off the big taffeta business. I never have the right clothes just, say, to go out to drinks. So I'll borrow a Betty Jackson jacket from Mum and put on stiletto heels which make me feel smart. Though I can't walk in heels very well any more. When I was seventeen I used to wear them with jeans. I'd never do that now. Overcoats have been another change: I used to wear huge men's overcoats. I don't any more, they look drab. I've got one from a friend that's a Chanel copy. I think my change in taste is slowing up. I hate to admit it but I was *so* affected by everyone else's taste at Oxford. I always imagined I was wearing the same as everyone, so I was very pleased if someone thought I had my own style.

'The big shock,' Miranda continues with a smile, 'was September 1985, the bank – working clothes. The working uniform one has to conform to. I knew you were meant to wear a suit: luckily I had a Jasper Conran one. You're meant to look neat and new and good: skirt and shirt, no very bright colours, never coloured tights – if you wear black ones, even, they think you're going on to a party after work... *I'm positively unhappy in working clothes*. I don't want to be that sort of smart, to look like an air hostess. And it's so important to get nice suits. The only ones I've ever found have been over two hundred and fifty pounds. I bought one for thirty pounds from Top Shop but never wore it. I suppose I should try Next, but I just don't like their things, and you can't get work suits in Joseph. I'll probably have to go to Harrods or Harvey Nichols in the end, as they have thousands to choose from.

'I've absolutely no buying plan. If I tend to find a jersey I like, say, I'll get it in several colours. I've got three with collars and buttons, and five pairs of leggings: I'll wear them till they get too revolting then go and buy another set. If I do get anything expensive I tend to think I'll look after it. But I never seem to be able to save for something expensive. Shops I go to are mostly Miss Selfridge, Joseph and Warehouse. I get shoes from Hobbs and Bertie. I

No minis allowed in the City: even among young *women a conventional uniform must be adhered to. Pinstripe suit by Jasper Conran – the sort of thing Miranda hates having to buy but admits she is now getting used to wearing.*

AKAI
JVC

MIRANDA BRETT

love those skating boots with little buttons – I long for millions of pairs of good quality shoes. I'm not a frilly person, though I used to be, so don't often go to Laura Ashley except to get the odd plain skirt or jersey. I tend to wear rather long things, meant to cover up my bottom. I long for an eighties figure, but I'm too fat. I like long tops, clothes that fit. I'd never wear an enormous, baggy T-shirt. My favourite fabrics are cotton jersey, lambswool, cashmere if I can afford it. And I secretly love velour, though I've yet to find anyone else who likes it. What I absolutely hate is anything shiny or synthetic. As for jewellery, I love *anything* from Butler & Wilson, but not big coloured stones and diamanté. I wouldn't want any real jewellery as we get burgled so often, and anyhow I have no longing for it *now*, though I might when I'm older.'

The person whose dressing Miranda most admires is Corinna Dundas, who is married to David Dundas, writer of pop songs and jingles. 'I really admire her. She dresses quite beautifully. She's got the Oxfam knack, and yet she spends a lot of money on clothes too. She always looks fantastic: I'd like to dress like her or Pamela Stephenson. I'd *love* to have the Oxfam knack myself, finding the right thing cheaply. But I haven't. I don't know how to describe it, but I'm conscious of never being up to scratch except perhaps at a proper party.'

Miranda's friends would probably be surprised at her never-mentioned despair about her own clothes, for her own style, no matter how agonizingly it may have been achieved, has always been

Miranda still can't quite kick the mini habit: this velvet dress was bought for her nineteenth birthday party from Photo in the King's Road – her only grand party dress, much loved, with a long life ahead.

admired for originality of colour combinations.

'I do like my clothes to be noticed,' she says, 'and for people to comment, full stop. If they're terrible, it's worth knowing. My younger sister Becca is marvellous about that. She's totally honest in her criticism. Whereas Mum, if I'm whingeing on, will always say whatever it is is fine, which is pointless. I dress mostly for myself and certainly a bit for my boyfriends, though I mind less and less if they don't like what I'm wearing. I know roughly about the actual fashion world because I do look at *Vogue*, though I'd never dream of wearing the sort of things they photograph. I'm a sucker for consumer clothes – I liked polo necks last year because they were in. The year before I wouldn't have been seen dead in one. On the whole I like the clothes people of my age wear. Right across the board I think they're well dressed. But as for Englishwomen in general . . . they're terrible, the women you see in the street and in Sainsbury's. Sometimes I sit on the tube wondering how on earth they could possibly have chosen. . . . All those ghastly fabrics, so often positively displeasing to the eye. And it's a matter of taste, not money.

'I have a sort of dream,' she laughs. 'What I'd really like to do would be to go to Harrods for an appointment with Mr X who'd found clothes for me and they would be *right*. But I haven't found a Mr X and I don't trust anyone I have found. Ideally I'd like to be smart but not serious and dull. I'd like *smart variations* of what I've already got . . . but then I suppose you can't have smart leggings.'

ELIZABETH SIEFF

WIFE, MOTHER AND HOSTESS

'My mother used to have to turn my father's old trousers into school skirts for us. They were beautifully done: I was very proud of them . . .'

For weddings and especial day time occasions: pink tweed and matching fur from Norman Hartnell – it also has a pink and grey silk shirt. Elizabeth much enjoys the occasional luxury of 'shopping' – though so mundane a word seems inappropriate to a visit to Bruton Street – at Hartnell. 'They take such trouble and have lovely big dressing rooms,' she says.

Elizabeth Sieff is married to Michael Sieff, former managing director and vice-chairman of Marks & Spencer, and a grandson of the founder. They live near Sunningdale and have two children. Elizabeth, a State Registered Nurse, used to work at Marks & Spencer herself: her job was to see to the well-being of retired staff. It was while working at Head Office that she met her husband. Since their marriage she has written a children's book, and devised an anthology called *Famous Firsts*, whose profits went to charity. She is continually, busily engaged in various works for charity, including fund-raising for the Windsor Hospice. Last year she was inspired to arrange a seminar for the associations that are responsible for children's welfare, to strengthen links and understanding between one organization and another in

FRENCH
FULL RED
WINE

the fight against child abuse. The project took a year to organize and the seminar will take place this autumn. Elizabeth tries to play tennis all year round, and goes to keep fit classes three times a week.

'My father was an actor and my mother a cook in the local factory so that she could be near home. Later she was a civil servant. I was one of four children, so things were tight - in fact so tight my mother used to have to mend our shoes and turn our father's old trousers into school skirts for us. They were beautifully done: I was very proud of them. I can remember as a child looking at lovely dresses in shop windows and knowing I could never have them. I can't remember having *anything* new. I wore my sister's hand-downs.'

The fact that she can have them now does not mean that Elizabeth spends a great deal of her time thinking about clothes. 'I've always been interested in them and think they *are* important, though I'm so busy they have to come second. I never *talk* about them to my friends. I make all decisions for myself. I'm not constantly on the look-out. I'm an erratic buyer. I suddenly realize that I must do something about my wardrobe, or get something for an occasion that's coming up: buying clothes really depends on what we're doing. I do pay a little bit of attention to fashion, though if it's something that's going to show up my worst feature, however fashionable, then I avoid it. Long may padded shoulders and long skirts go on. I love them! I also love coloured tights and, influenced by the children's nanny, I wear masses of them. My husband hates them. To be honest, he doesn't like me to be too way out. He still likes the old shirtwaister and flesh-coloured tights. Sometimes he doesn't like the clothes I've chosen at first, but when he gets used to them he may mellow a little. I try hard to please him and my clothes are really a combination of what Michael likes and what I like.

'I don't think my tastes have changed greatly. I've always liked the same sort of things – I suppose you could call me a variation on the classical dresser: slightly romantic and floaty sometimes. I like a lot of variety, and I wear my hair up or down, or tie scarves in it, to go with what I'm wearing. I love bright colours, particularly red. I also like white and pale pink, but I'm not keen on sludge or grey. I love pure materials – silk, velvet, real wool – though not linen. It's so impractical for travelling. The straight look I tend to go in for. I've got several straight skirts, some with kick-pleats, and four dresses that hang straight from the shoulder. Well, I like trying for the tall slim, elegant classic line: that makes me feel good. I'm very fond of batwing sleeves, and suits with a hat to match – hats I particularly love. I wish they'd really come into fashion. I've an amazing white one, three years old. I put ribbons and streamers and bows on to it, and it goes to weddings and Ascot. I keep things for ages if I like them.

'I'm happy to try everything. The only things I'm not keen on are high, round necks, full culottes or skirts, short skirts, waists and frumpy cardigans. For every day most of my clothes are from Marks & Spencer – usually from the Staines or Marble Arch branches. For winter I buy those lovely soft thick tights; underwear, jerseys, skirts, blouses, mackintoshes and coats. I'm very thrilled and proud of Marks & Spencer: and by and large I think they're fabulous, wonderful value. If there *are* any criticisms, and obviously every single department can't be at the same level all the time, I put them to my husband who passes them on. I also go to a couple of local shops. One is Jillifar, a small boutique in Chobham. They know my taste and I get nice woollen things there. The other is Casita Ray in Sunningdale, which has a good range of medium to high price things. In London, I love Harvey Nichols. But for something *very* special, I go to Hartnell. They take such trouble. Lisette, who always helps me, is so enthusiastic. They have lovely big dressing-rooms – it just makes you feel good, going there. Then a friend told me about Benny Ong, who supplies clothes for Harrods and so on. He has the most unusual clothes, a really exciting range. I've got a marvellous white silk sort of nineteen-twenties evening dress from him.'

Elizabeth's favourite boots and shoes are by Rayne. 'I love high heels for evening,' she says, 'though my daily runaround shoes are from Marks

& Spencer with tiny stumpy heels – so comfortable and only ten pounds. I'm fond of high-fashion costume jewellery – the new M & S collection is very exciting, incidentally. I've very little real jewellery, but what I have I mix with semi-precious stones – coral and jade are my favourites. I also mix it with costume jewellery.'

There's a touch of *Dynasty* about Mrs Sieff's wardrobe, with its floor-to-ceiling glass doors that stretch the length of the bedroom. Inside sparkles a democratic mixture of Hartnell, Ong and Marks & Spencer. She has several cloaks, for both day and evening, and a preponderance of evening dresses, many of which are very old favourites. 'I really enjoy dressing up for the evening. Round here people change for dinner parties and are quite smart, though the women always tend to wear short dresses.'

Englishwomen, in general, Elizabeth feels, are better than they used to be at putting things together. 'Well, they really ought to be – there's so much to choose from. On *sporting* occasions they look very good – their best, I think. And as a matter of fact the mothers outside my children's school dress pretty well. But on the whole, the French are far chic-er, chiefly because of the way they *walk*. The English aren't nearly so good at moving, showing themselves off to their best. I loved the way the late Princess Grace used to dress, and I also think the Duchess of Kent and the Duchess of Argyll are always beautifully turned out.'

Elizabeth Sieff herself likes to be complimented on her appearance. 'Any appreciation makes me feel wonderful. I say to Michael: "Do I look nice?" and he says "Yes, but you always look nice." I envy the way the *young* dress today,' she goes on, 'the freedom of it all. What I really look forward to is when I can wear more lovely suede and leather. But that'll have to be when the children are grown up, and the dogs are better behaved.'

Previous page: For more ordinary days Elizabeth is loyal to her husband's firm and buys a quantity of separates, mackintoshes, underclothes, shoes and jerseys from them. Everything here is from Marks & Spencer. The scarlet shirt, reminiscent of Jasper Conran, is 'pretend silk'. The black beads, from 'the very exciting collection of fashion jewellery' were available last year.

JENNY ANDERSON, MBE

— MANAGER OF SPECIAL SERVICES, GROUND OPERATIONS, — BRITISH AIRWAYS

'Matching, matching, matching – the together look is what I like. I'm sure that's frightfully old hat, but I do it relentlessly.'

One of Jenny Anderson's more formal executive suits, the sort of thing she would choose on days VIPs must be ushered through Heathrow. It's pure wool, by Daks, from Simpsons. Red silk shirt from Jaeger, Gucci bag.

Jenny Anderson has spent her whole career working for British Airways. She began by typing passenger lists, graduated to reservations and in 1966 became their first female passenger officer, which she feels was something of an *avant garde* gesture on the part of the airline. 'It was a hard school to be proved in,' she said. But proved she was, and progressed up the hierarchy until, two years ago, she was not only made Manager of Special Services, but was also awarded the MBE. Jenny has a staff of thirty-five working under her: they are responsible for making sure that VIPs have an easy and protected arrival or departure through the airport. If there are any problems, Special Services deals with them in advance so that the passenger may never even realize there was a problem. VIPs include ill children travelling to Disneyland, members of the Royal Family and heads of state, and often completely unknown travellers such as the old couple who, celebrating their sixtieth wedding anniversary, found British Airways had provided champagne in the executive lounge.... Jenny's 'laughingly called' office hours are nine o'clock till seven or often later, and she lives near Heathrow. All her energies are expended at work and much of her social life is connected with the airline. When there is time to relax she likes to go to the theatre or watch television and is 'an enthusiastic reader of light fiction'.

'My aim is to achieve a level of elegance on a shoestring,' says Jenny. 'I don't get a clothes allowance from the airline, but I like to create a good image for them, always to be clean and well groomed. My clothes have to be becoming to me, and I have to look appropriately dressed. Generally speaking, I dress formally – in fact, I have few casual clothes. Often at the end of the day there's a cocktail party or reception I have to attend – I *always* change, either at the office or home. I haven't any cocktail dresses *per se*, but I put on something lightweight, normally covered up, but a dressy dress that would be out of place if I wore it all day. I don't care too much about *fashion*: it seems all to be designed for the young these days. But I try to pick up fashionable colours. More than *Vogue*, I trust my sister, who works in a boutique, to tell me what's coming. But I do look at fashion magazines and find a few suitable things. I like the white collars and the fuchsia pink tights that were all the thing in the summer of 1985 – I wore pink tights with a pale grey suede suit and a fuchsia pink blouse.

'I would say I'm a romantic-classical dresser. Tailored clothes are what I love, suits, and three-piece outfits, never odd separates. Matching, matching, matching – the together look is what I like. I'm sure that's frightfully old hat, but I do it relentlessly. I always try to match my tights to my skirt and I usually match even my jewellery – semi-precious things such as rose quartz, which I wear with pink, serpentine beads with soft green and so on. Bags and shoes *invariably* match. I have fifty-three pairs of shoes – I will carry one about endlessly searching for the right colour bag to go with it. Once I bought a pair of shoes for £8 and eventually found a matching bag for £60. That's the sort of silly thing I'll do. I think accessories are so important.'

On the occasion of her investiture at Buckingham Palace in 1984, Jenny chose a lilac wool-lace suit with a matching coat. To make assurance double sure on the day, she equipped herself with *five* sets of accessories from which to choose: lilac, cream, navy, light-brown and claret. She wanted the comfort of choice. 'I'm a conscientious shopper,' she confesses. 'I enjoy it immensely though I have very little time. Occasionally I have to take a day off to go to London. I like to have a mix of clothes, a base of good quality classics like a Daks suit from Simpsons, and a standby navy suit and white blouse, and then I go to Marks & Spencer for more fashionable things in the season's colours – shirts, sweaters, summer frocks – which I'll discard at the end of the season. For evening things I go to John Lewis' sales, and Dickins & Jones. I like mid-price designer things – I love Frank Usher, for instance, and I yearn for a Bruce Oldfield, Jean Muir or Oleg Cassini. What I would really adore would be a Chanel suit. But I haven't that sort of money. I've got good stuff from Debenhams including a beautiful three-piece: they're good at what I call rather Jewish wedding outfits. Once a year I shop in San Francisco. The chain shops out there have a huge range, and Loehmans is wonderful - it stocks all designer goods of the previous season, with their labels cut out, and dramatically reduced.

'To my great regret I can't wear black as I look bereaved, and I don't find grey very becoming. I adore scarlet and turquoise and blue and green. For summer I'm perfectly happy with synthetic materials that will drip dry. I don't wear much real cotton or silk – they need too much care. But I love

real wool. I don't like an Empire line – I like waists and long floaty skirts, but I can't wear *them*. I'm far, far too old. I keep them for lingerie. I never feel I look good in trousers, though I do the housework in them. I've a cashmere coat, two mink coats and a mink jacket – I regard them all as *investments*. I keep things a very long time if I like them. Hats I only wear at weddings, hats with brims. I think veils and feathers are too twee by half, though I can take a flower to soften a hat in summer. Because of my romantic side I sometimes imagine myself in Henley-type clothes – straw hat with a ribbon down the back. I have a frilly soul but unfortunately I don't lead a romantic-type life. In reality clothes are all part of my work and it's important to me that I feel good, and comfortable. I have to do so much walking about the airport every day that I find sandals more comfortable than court shoes – what I adore most are T-bars. I'm constantly moving from hot to cold, so always need a jacket rather than a woolly.'

One of the bonuses of Jenny's job is that she is in permanent contact with the travelling public, which she enjoys. From her vantage point, she is able to observe the standard of travelling clothes. 'I'm always pleased to see women beautifully groomed, and put together, and there *are* quite a few of them travelling on Concorde. I think it's a matter of taste: money has nothing much to do with it, though I suppose it is easier with money. As for the passengers in general . . . well. What I really loathe is the jeans brigade, especially on older people. The art of travelling is to wear something comfortable, I know, but why can't people change into something smart before they land? Not many people do that. Standards of English dressing aren't high, though it depends which circles you mix in. My friends are always well-groomed and my mother, at eighty-four, is still in red trousers and beaded cashmere, so perhaps it's all to do with conditioning, what's instilled into you. When my sister and I were little we were always spotless. But I don't think dressing is an English trend. I think those intellectual women who somehow think they are above trying with their appearance are very arrogant. Some women may enjoy making an effort in the evening, but at tennis dances and Ladies' Nights the apparel you see defies description. I think the Duchess of Kent always looks superb, beautifully put together. Among the stars and actresses I see no one comes to mind. The Queen Mother is the epitome of everything: she always looks divine.

'I love compliments,' Jenny goes on, 'but they're no good unless *I* think I'm looking okay. (If my tights don't exactly match my suit, for instance, I'm inhibited.) I find that the people who flatter one most are invariably women. They recognize *quality* but don't pass any comment on more ordinary things. The nicest compliment I ever had was from Georgina Andrews (Anthony Andrews' wife) when she said what a lovely dress I was wearing. I dress to look my best for the airline, and of course for *me*. I believe it's a compliment to those you're with to look your best. With all due modesty I could say I was known for my clothes. Sometimes, I look in the mirror and think: you're not going to look *better*, at your age.'

Previous page: From Casual Corner in San Francisco, a favourite for 'summer cocktails'. An example of Jenny's love of things matching, though the top is not pleated. Material? 'Some mixture, I suspect.'

VIVIAN STUART

HISTORICAL NOVELIST

'I've worn trousers all my life. If I have to dress up at night I wear a man's dinner-jacket, but with a string of pearls to show I'm a woman.'

Vivian Stuart has written eighty books – historical and romantic novels and two naval histories – and, with Denise Robbins, she founded the Romantic Novelists' Association. She is currently working on a 'factional' history of Australia: each of the eighty titles so far has sold a million copies in paperback in America. Born in Rangoon – her grandfather founded the Burmah Oil Company – Vivian Stuart was brought up in England and Scotland. She went to India and married a Hungarian, became a refugee in 1938 and followed

Lest anyone should think that Vivian Stuart is a man in this severe get-up of catalogue jacket, trousers and black shirt, she makes a single concession – a pearl necklace, to indicate she has had four husbands and has nine grandchildren.

him to Australia. Now, in her house just outside York, she writes seven days a week, six to seven hours a day, typing about fifteen hundred words a day with one finger. She was chairman of the Writers' Summer School at Swanick for six years, and is now on the committee, and she attends the annual Writers' Conference at Scarborough every year. She has five children and nine grandchildren, and has been a widow for the past two years.

'Frankly, there's not much time to think about clothes,' says Vivian. 'I've no time to go shopping – haven't been into a shop in York for twelve years. So I order things from catalogues. I suddenly get a thing that I haven't got enough anoraks, so I do a bust and order three. I decide two make me look fat and give them to the Salvation Army. Kay's catalogue is best. Shoes, socks, I get everything from them. Sometimes five trouser suits at once: they're what I wear practically all the time.

'I've always loved men's clothes,' Vivian explains. 'When I was a child I prayed at night I'd wake up a boy: I wanted to join the Army. When I was twelve my father bought me a beautiful Guard's uniform. I went to a fancy-dress party in it and a girl fell for me – she couldn't tell I was a girl. I was mad for horses as a child and lived in breeches and boots, and shorts for tennis. I did get into the Army eventually – I was in Burma during the war, and of course wore trousers or shorts then all the time. I do *own* some skirts, and occasionally put them on: but then I go back to my trousers.

'I pay no attention to what's in fashion at all. I know what suits me and what doesn't. I try not to be mutton dressed as lamb, and I try to dress with a certain dignity and a sense of humour. I'm a warm and comfortable dresser in winter, and a cool dresser in summer. Horrors, to me, are materials that crush. I want to arrive uncrushed, so I do like synthetics, and acrylic jumpers. Lambswool is really my favourite, but acrylic won't put me off. I can't bear stockings, always wear socks. I like comfortable shoes, never high heels – I'd fall on my nose. I don't like yellow, but reds and bright colours, and blues and mauves merging, and black. Since I've been getting older I like to wear a scarf round my neck with literally everything. I used to love fishing hats, and once I had a very nice felt top hat. Then I had to spend two years with a patch over my eye so couldn't wear any hat. Now, I have a fur hat to go with a good wool coat when it's very cold.

'I've always had the same feelings – I'm happiest in anoraks and trousers, and open-necked sweaters or high-necked jerseys. I dress entirely for myself – my husbands never used to criticize. In fact, I do order frocks, occasionally, from the catalogue, but they *must* have pockets for cigarettes. I will buy a frock even if I don't like it so long as it has pockets. I've got one posh diaphanous thing called a lounging-gown or something – it rolls into a ball and comes out uncrushed. Useful for Australia when I go and visit my son. I have the same trouser suit – some sort of synthetic stuff – in five colours. I don't wear the orange one here, but I do in Australia.

'My sister,' Vivian goes on, 'is a great dresser and thinks my things are terrible. But then I don't care for compliments about my clothes – I'm more interested that people should like my books. My youngest grand-daughter has spikey hair and white stockings – I think it's awful, not attractive or anything. Her sister is more restrained but dresses in a nineteen-twenties style. I like kids in jeans and colourful shirts. The middle-aged Englishwoman is inclined to be dowdy, not adventurous at all. But I do admire the Princess of Wales – she looks nice in anything. Absolutely lovely. And Selina Scott's top half I rather like. She had a nice buff shirt, double-breasted with pockets – must get one. I never generally notice what people wear. I'm more interested in their eyes or their smile. If I do notice, it's because it has pockets. I've an incredible anorak

Vivian Stuart has two choices for the evening: the most dramatic is the dashing, knights-of-old combination of tartan knickerbockers, velvet jacket and lace jabot. They all come from a mail order catalogue, even the shoes.

that has *five* pockets . . . '

For the 'posh night' at the Summer School at Swanick, Vivian puts on a frock, while for the evening at the Writers' Conference at Scarborough she presents an arresting sight, handsome and dignified, in tartan knickerbockers, velvet jacket and stock. 'The whole outfit came from a catalogue,' she laughs, 'including the shoes. When my father, Sir Campbell Kirkman Finlay, was knighted, and had Court dress, I inherited the shoes. I took off the buckles, and put them on my catalogue ones. The tartan knickerbockers started off as trousers but my dressmaker altered them. The green velvet waistcoat and black velvet jacket are, surprisingly, women's.'

An alternative evening dress, which she might wear to a Burma Star dinner in York, is a man's white dinner-jacket, black trousers, and black shirt. . . . 'I would say I was an eccentric dresser,' says Vivian, 'I dress entirely for myself, but I don't want to be taken for a man. So when I wear my dinner-jacket I make one concession and wear a single string of pearls.' She smiles. 'Well, I've had four husbands and I have nine grandchildren. Pearls, I hope, indicate this.'

TESSANNA HOARE

PAINTER

'I think elegance *is what's lacking in my generation. To people of my age the idea of dressing up has gone out of the window. Given the chance,* I do *enjoy it: but it has to be for some formal Lady So and So invites you to . . .'*

Tessanna's mother says her feminine side has 'gone out of the window . . .'. Typical of her biking clothes: black skiing trousers from a secondhand shop, green cycle top worn over a black suede sixties vest belonging to her brother. Brown walking shoes from the Natural Shoe Store.

Tessanna Hoare began by training as a dancer at the Royal Ballet School, then went on to The Place. She spent two years working as a waitress and modelling for Laura Ashley, and four years at art schools – The City and Guild and Camberwell. When she left in the summer of '85, she almost sold out at her degree show. She has had three shows since then, her latest at L'Escargot Restaurant (where she was once a waitress). Besides painting, Tessanna designs and makes tapestries, designs and paints theatrical sets, makes the-

atrical costumes and also does some teaching. A 'bike fanatic' who enjoys 'bike messaging' when she wants to earn some extra money, she moved into a studio off the Portobello Road earlier this year.

'I'm very lucky,' says Tessanna, 'living in this part of Notting Hill. It's like the Left Bank in Paris. I'm surrounded by painters, sculptors, dress designers, jewellers, furniture designers – we're all *working*. If you're expressing yourself through your work, then you don't need to think much about clothes. They come secondary. It's only if you have a boring job and have to look pretty behind a desk all day that your clothes become important as a means of expression. As it is, I find myself wanting to make do with less and less. I just don't want to go to shops – it's all so over the top, so commercial and you come out looking stereo-typed. It's not that I would mind looking like everyone else, but the excitement wouldn't be there. I do love clothes, in fact, but wearing nice ones is completely impractical. I save them for special occasions and basically live in jeans, although I have good jackets I can put over them to go out, and look quite smart. Mum says, "Your feminine side has gone out of the window." But I haven't a feminine figure: I'm very boyish.

'What I care about in clothes,' she says, 'is the making, the quality, the craftsmanship. That's why I buy old things in markets – the Portobello Road and Brick Lane. I do a market about once every three months. I spend very little – I don't *need* clothes, so I only buy something for the thing itself. If I make a mistake I sell it again. I never stick to one style – all sorts of different styles suit me, so I buy a one-off if it catches the eye. I love French and Spanish style things, and primitives, and dresses and jackets from all eras. If I had a lot of money I would get friends to design and make things for me, and I wouldn't mind things from Kenzo, my favourite designer. I do have one dress by him, which my sister gave me. The only thing I'm tempted by in shops are workmen's clothes, fishermen's jumpers and trousers, for instance. I buy them in Norfolk. I think working clothes are the best-designed and best-made things you can find.'

Tessanna describes herself as a 'colour dresser'. 'That's what my friends pick up,' she says. 'They say "This girl's good with colour." (I do enjoy compliments from friends and others, though I like it better if they say *you're* looking nice, rather than just the clothes.) What I love are *sea* colours – blues, greens, turquoises, colours in their purest form. I've a lovely fifties silk two-piece in turquoise silk. I also love lemon yellow. I hate burgundy, it looks dreadful on me I reckon, and I've never gone for chocolate brown though I probably could after a while . . . I love real silks, mostly Thai and shot silk. When I was in Thailand I found beautiful silks for scarves. I wear synthetics on the bike and for practicalities, but I'm a bit of a snob about them. I can see their beauty – last for ages, easy to wash and so on, but I don't like them next to the skin. Rayon isn't *too* bad, but Crimplene is awful. I can't remember when I last wore tights. I hate the feel of them. I wear socks with everything, even when I dress up. My shoes are mainly practical American shoes and flat pumps – I've never been known to wear high heels. I have a few hats and especially like a traditional Russian one of grey rabbit fur. As for jumpers – I'm a jumper fanatic.' A jade-green jumper Tessanna had recently bought at a sale was the first new thing she had had for as long as she could remember. 'I don't mind ruining it,' she says, 'but I would mind ruining my favourite orange and yellow and blue one knitted by some old dear that I picked up in a market. I'm also an *accessory* fanatic: scarves and ribbons I love. I spend so much time painting the only way I can sparkle in the daytime is to put on scarves and ribbons. . . . Jewellery I used to love, but I'm keeping it at the back at the moment. I tend to be given real things by friends – I don't like tacky jewellery.

'I think my tastes are pretty constant and always have been,' Tessanna goes on,' though occasionally I'll go off on some reaction. I automatically follow what's going on in fashion because of my sister (Sarajane Hoare, fashion editor of the *Observer*) but I try not to take any notice. I suppose I do pick something up, whatever. For instance I never used to wear black, then fashion went black and I got into it, though it didn't happen very consciously and it's gone now, that phase. Maybe,' she adds, 'people started wearing so much black so that *faces* could be focused on. Abroad, it's the faces of the working people you're drawn to, all in their uniform black. Here, I'm drawn to colour, looking at other people's clothes, but I prefer to look straight at their eyes.

'The main thing about my own dressing is that I should be comfortable. I rarely wear skirts, and don't like frills or waisted things, nothing too busy. What I would love,' she says, 'is to be able to dress up and not feel stupid. But the only time you can really do that is at clubs, where everyone is so busy trying to be individual they go way over the top. I think it's a pity there's not more *exuberance* in the way Englishwomen dress: you very rarely see those who are a pleasure to look at. But perhaps it's because of the whole ordeal of shopping. Standards have gone up a lot – Marks & Spencer is quite stylish, but it's all so over the top people find they don't want anything: they can't face the ordeal of buying. Among my own friends the general feeling for clothes is clean and understated: I *don't* think they're scruffy at twenty-four, though perhaps they are at nineteen. I think my sister Sarajane is one of the best-dressed girls in London. She is so compact, so behind it. I don't think famous people dress very well. I can't think of any I admire.

'I think *elegance* is what's lacking in my generation. To people of my age the idea of dressing up has mostly gone out of the window. Given the chance, I *do* enjoy it. But it has to be for some formal *Lady So and So invites you to* . . . occasion. Dad loves it when we dress up. It's very rare, but it's a real pleasure – the waking up and inventing the outfit, which takes a long time, and then the bath beforehand. What I love seeing is everyone in conventional white English balldresses. Two years ago the Ashley family gave a ball in France. It was marvellous. Everyone had really tried.' Tessanna glances up at a photograph on her wall of ballet dancers in identical white dresses swirled by their partners in uniform black. 'That's the sort of thing I mean: nothing higgledy-piggledy, no one trying to be individual. It's fantastic when you *find a whole room* of people is a feast for the eye: but it's a very rare sight.'

Previous page: For occasional grand evenings: petrol blue crushed velvet skirt and Chinese silk shirt, both from a secondhand shop. White fur cape from the Portobello Road market. Shoes from Hobbs – 'never high heels'.

JANET GLEAVE

PROFESSIONAL BALLROOM DANCING CHAMPION

'I thought those great net things we used to wear were horrible. I hated walking down a High Street with a bustle of fluorescent tulle on my back. . . .'

Gaberdine trousers by Jaeger worn with a variety of jerseys: the most practical uniform for teaching dancing, which Janet and her husband do several times a week.

Janet Gleave met her husband, Richard, at dancing school, and they have been married twenty-three years. Specialists in old-fashioned dances such as the waltz, foxtrot, quickstep and tango, they held the title of World Professional Ballroom Dance Champions eight years running – they retired in 1981, the year of their eighth win. They now coach five days a week at their studio in Kingston, and are much in demand both to present trophies and give demonstrations at various functions all over the country. They still travel to many parts of the world, though not quite so frequently as in their competition days, and they live in a house near Windsor.

'The dancing world has fashions of its own,' says Mrs Gleave. 'I thought those great net things we used to wear were terrible. I hated walking down a

High Street with a bustle of fluorescent tulle on my back. . . . Luckily, they died out about four or five years ago. Now, though they took a while to catch on, the dresses are lovely and flowing. The trend at the moment is for pleated georgette with feathers on the hem. I used to have a new dress for every major competition – six or seven a year. Now, I only have one, which I wear for demonstrations.' This one dress, which cost hundreds of pounds, was made for Mrs Gleave by a dressmaker in Japan. In the dance fashion of the moment, it is made of pleated georgette – a polyester georgette chiffon, to be precise – of sunny yellow, ostrich feathers wisping round the hem and a thick encrustation of sequins, bugle beads and seed pearls on the bodice, which splatter more thinly over the skirt. Although the sequins go round the arms to give the impression of looking off-the-shoulder, for practical purposes a thin layer of flesh-coloured nylon going *over* the shoulders keeps the top from slipping during a quickstep. 'The nylon doesn't show under the lights,' explains Mrs Gleave. Such dresses, of course, are made to look their best under powerful lights and on television – hence the ubiquitous sparkle and waving feathers. Essential to the whole structure are the myriad organza petticoats. 'A dress like this costs a fortune,' says Mrs Gleave, 'all hand sewn, every bead put on by hand. *Hours* of work. When one is finished I pass it on. Most of us get material from Nevilles Textiles in Nottingham, which specializes in dance-dress stuff: I study their catalogue and their shade card. They tend to supply normal colours for dance dresses. Yellow and white are my own favourites. I go along with the pleated georgette fashion as it's the best material for what you want. It's got weight, yet it's soft. I like a waist, then flatness to the hip, so the whole dress is a shape that gradually gets fuller at the bottom. The skirt always has to be the same as it has to dance.

'More ordinary dresses,' Mrs Gleave goes on, 'are my working clothes. I have to have quite a few for judging competitions and presenting medals, which we do all round the world. I get them in London stores – Harrods has a good selection, and Dickins & Jones. But I also find dresses in Cognitos in Windsor: they have a variety of things by people like DP Fashions and Frank Usher, who I like. I buy whatever takes my eye, but what I usually look for is something with a high neckline and long, drapey sleeves. I've made quite a few things for myself: on the whole I go for plain materials.'

For teaching her dancing classes Mrs Gleave wears trousers and shirts or jerseys. 'Jaeger trousers seem to fit me,' she says. 'I tend to go for the same colour schemes – black and white and creams I like, and I'm very fond of electric blue. My shoes for teaching are flattish and not very elegant, but I have to demonstrate the man's action as well as the woman's, so I can't wear high heels.' Even when she is not teaching, Mrs Gleave prefers separates – matching or contrasting – to dresses and suits. 'I don't get much opportunity to wear that sort of thing,' she says, 'and anyhow I usually find the waist and belts on dresses are too high for me. I'd describe myself as a smart-casual dresser, not a T-shirt sort of casual. I like classical things: I've similar taste, I think, to the Princess of Wales.'

There is not much time for shopping in Mrs Gleave's busy life, but she managers to 'buzz in and out,' at lunch-time in Kingston, to Bentalls and British Home Stores for separates. 'I love shoes,' she says, 'but often can't find what I want. I go to Bruno Magli and Kurt Geiger – I've got very expensive tastes, I'm afraid, but they do fit my feet. I like high heels and nice shaped heels – high and low. The problem is getting evening shoes. I saw some lovely black suede ones with a bit of gold up the back, but they were three hundred pounds. I found some pretty slippers in France, satin mules with diamanté. I bought two pairs and wear them as evening shoes for judging. For actual dancing my shoes come from Supadance. They're normally satin and can be dyed any colour to match your dress. Then I stick on jewels or a motif to match as well. They have special non-slip soles.

'I tend to shop spontaneously,' goes on Mrs Gleave. 'I go out with an idea of what I want, and try to find it, but if I suddenly see something I like I'll buy it. I do take an interest in what's in fashion and look at magazines, but I only wear what I like,

JANET GLEAVE

what suits me. My tastes have changed a bit: I used to like the slim-tailored look – I still do, but nowadays I've gone a bit more baggy and I love batwing sleeves. I like real materials, but synthetics are better for travelling. What I dislike is the casual scruffy look, and I'm not too keen on layers, either. I like frills but they don't suit me, and I don't like pink. I love hats but I haven't the confidence to wear them. I have three, but they just sit in the cupboard.

'I like coloured tights on others, but only wear flesh-coloured or black myself. I'm a bit conservative like that. I don't wear much jewellery except for earrings. In real life, as it were, I have a pair of real pearl and gold ones, and a real gold pendant. For dancing, you can't really wear anything except earrings: sometimes I make my own – buy clasps and sew on a few sequins. Dancers,' she adds, 'make enormous efforts about their appearances for competitions and I would say that on the whole they're very fashion-conscious and well-dressed lot. They get used to thinking about shapes, and putting colours together, and do it well. But the Englishwoman in general lacks flair, although I would say her dressing is getting a little better. The Princess of Wales I admire enormously: she has terrific style, is very natural. And Margot Fonteyn: she's always very elegant.

'I just slung it together,' says Janet Gleave of this bronze sequin evening dress. It's very simple, one of my evening work dresses which I wear to give away trophies or judge competitions.'

'My husband's taste is more or less the same as mine. If he doesn't like something, he tells me, and if it's a dress for a job then I'll listen and probably change. But really I dress for myself. I do get compliments, from both men and women, which is nice. I love clothes and spend quite a lot of time thinking about them: I enjoy the process of getting up and planning what to wear every day, and I enjoy buying new things, though they don't have to be expensive. To choose the right clothes for judging takes a lot of thought and planning. I tend to think in advance what I'm going to wear, and make sure it's clean. The actual changing for the evening is often a bit of a rush.'

Despite the pristine appearance of Mrs Gleave's yellow dance dress, the time is approaching for a new one. She has in mind candy-pink pleated georgette chiffon decorated with silver beads and rhinestones, and more feathers on the hem. This one she will make herself. 'The careful planning and thinking, and then seeing it materialize, is very satisfying,' she says. 'Each dance dress is a work of art.'

The tulle bundles of yesteryear quite gone, this is the new look in professional ballroom dance dresses: pleated georgette chiffon, ubiquitous glitter and feathered hems that swirl out to 'dance'. This one was made by a dressmaker in Japan and cost over £1,000.

SUSIE FRANKLYN

HOUSEWIFE MARRIED TO THE ACTOR, WILLIAM FRANKLYN

'I like to look pretty yummy, to be noticed . . . Mentally, clotheswise, I'd like to be the eternal mistress.'

Susie's £15 Oxfam dinner jacket apparently 'belonged to a man of importance'. Beneath it, a pale pink pin-tucked shirt, copied from a man's, from Hennes. The sparkling cross 'adds a touch of Marlene Dietrich.'

Mrs Susie Franklyn, once an actress herself, is now a full-time wife and mother, 'incredibly involved' in her family's activities. Her husband writes as well as acts, and she herself, having been born in China, would one day like to organize the dramatization of the run-up to the invasion of Hong Kong in 1942. As it is, she has started a novel. The Franklyns have two daughters and live in a pretty house in Putney where, she says, her days are 'absolutely full'.

SUSIE FRANKLYN

'I am,' she declares, 'a clothes freak. This may have come about because for a long time I lived in extreme poverty. At boarding school I was the fat girl in the gravy-stained blazer buttoned to cover up my enormous tits. I used to think then: one day I'll have a go at looking better. After art school – where I learned dress-designing and making – and a typing course, I was allowed to go to drama school. At that time I was into duffle coats and hooped earrings, and went on to felt skirts and elastic belts. After dieting – I never got fat again – things began to improve. I got a black barathea suit with a velvet collar and nipped-in waist, and felt pretty A1 in it. It became my interview suit. I also had an emerald-green crossover blouse with a bow at the back. By the mid-sixties, when I was living in a very West Side Story sort of place off the King's Road, I was getting tarted up for the evenings, high heels clattering down the steps. That was the time I began paying real attention to clothes.'

Now, Susie Franklyn likes to be glamorous every day. On a September afternoon she wears a Persil-white boilersuit, scarlet high-heeled shoes and a diamanté necklace – second outfit for the day, for she puts on a lesser boilersuit for early morning housework. 'Susie,' says her husband, 'is an *aficionado* of gear. She has her own style, there's no word for it. She's a bit like lightning,' he adds, fondly, 'she tries not to strike more than once in the same thing.'

Her husband may have no word to describe his wife's style of dressing: Susie herself has no difficulty in explaining it. 'I'm an *impact* dresser,' she says, 'rather theatrical, I suppose. I once met a lady who was a brown person. I'm opposite to that. I like sporty clothes, but when I go out in the evening I like the works. I like to look pretty yummy, to be noticed. My husband is very appreciative of how I look. That keeps me on the ball. I don't think I make many mistakes.'

For evening dressing up, Susie is particularly fond of the sparkle of sequins: she has various sequinned tops, which she thinks very feminine but, in contrast, she also wears men's dinner jackets at night with exotic shirts. One of her favourite long dresses, which she still wears, is her 'traffic lights' dress (stripes of bright colours) from Ossie Clark in the seventies. 'It goes on because it's a classic,' she says. 'But I don't ever like to look completely out of fashion. I've a cupboard full of late sixties, early seventies dresses by Gina Fratini and so on, but I'd never wear them now. I don't exactly collect clothes,' she adds, 'but I do acquire them.'

The extent of her acquiring meant that when the Franklyns moved to Putney Susie had to hire a special van just for her clothes. '*And* Bill's,' she smiles. 'I do have an awful lot, I admit, but on the other hand I *don't* spend a fortune. I'm quite economical in some ways. For instance, there's a wonderful man's tailor in Putney called Dimi Major, who used to be with Doug Haywood. He altered and cut down two of Bill's old mid-sixties suits for me: and my very good dinner jacket comes from Oxfam. I either like very tailored or very feminine things. I wear boilersuits a lot in the day – I have to be practical. I've four lined cotton ones from Jigsaw in different colours. If I'm going out to lunch I might put on one of the trouser suits. I was the original mini skirt person, having reasonable legs. Now, I wear mid-calf skirts. Sometimes, I like a V-neck with a bit of cleavage. But really, separates, jackets with broad shoulders and clear lines – the Café de Paris look, that's me. In summer I wear a lot of Bermuda shorts. We often go to Italy, so I buy sporty clothes there, and bikinis. They're *much* better cut than here, especially if you have a big bust. If I could afford it I'd buy most of my clothes in Italy. Every year I'm stunned by how gorgeous the Italians look compared with the English. They have a way of putting things together – wonderful colours, lovely belts and gold jewellery.

'I'm an impulse buyer,' she goes on, 'though there's nothing willy-nilly about my wardrobe. I think about it all quite hard. I get up early in the morning and dress in the dark, but even then I don't just fling on anything. I give it thought, try to look reasonable for breakfast. I do the housework, then bath and change mid-morning. As for buying, I'm not snobbish, not a label collector. I just can't resist quantity. It's a bit like a hire shop here.'

Susie Franklyn

'Other people's friends come and look at their libraries,' puts in Mr Franklyn, 'Susie's come to look at her wardrobe.'

'But I'm not extravagant,' insists Susie. 'I buy a lot from a nearby second-hand shop. The white boilersuit, good as new, by Gloria Vanderbilt, I got for a song. I have a dressmaker round the corner so I often design or find a pattern and adapt it. I never make something entirely myself because I'm such a fusspot. But I do alter. I'm always altering, bringing things up to date. The sort of more expensive shop I like usually has an Italian influence – Benetton for jerseys, for instance. I like the King's Road, though I practically never go somewhere like South Molton Street. I'm too much of a fusspot to get underwear from Marks & Spencer. America has the best, and I like French bras. I sometimes find things at Bonham's sales, like my wonderful nineteen-thirties fur coat with *Gloria* embroidered on it, and diamanté buttons. I like to think it belonged to Gloria Vanderbilt. I get lovely padded jackets, very cheap from Putney Market, and if I'm lucky I find beautiful antique clothes from Rhoda Valentine. But I very seldom have a preconceived idea of what I want. Sometimes I go to acres of shops without finding a thing.'

Colours Susie never wears are brown, orange, sage, mustard, plum – all earth colours. Her favourites are black, white, grey ('even though I'm over thirty'), red, blue, violet, and, very occasionally, yellow. These colours she finds in 'all the lovely materials – cotton, wool, silk, cashmere, velvet. I used to wear velvet breeches because I loved the Little Lord Fauntleroy look of a few years ago.' As a

Another of Susie's bargains for the evening – a Janice Wainwright dress found in a secondhand shop in Putney for £30 instead of £300 . . . . 'The neck is ruched in a very unusual way,' she says.

young actress, to make a little extra money, she sometimes modelled fur coats at Swan & Edgar. 'I felt fabulous in them, like Cinderella in someone else's clothes. Fox is my favourite – well, foxes *are* hooligans – they killed all our chickens. So I have fox furs.' She is very fond of belts, too, and has built up an 'incredible collection' as she has of shoes. 'I've *dozens* of them in all different colours. When they go out of fashion I chuck them out. A lot of them are baseball boots, gym shoes, or jazz shoes in lovely colours, and masses of high heels and sandally shoes for evening. I *always* have a bag to match my shoes.'

Susie's dislikes are very clear. 'What I'm *not* is one of those twinset, pearls and pleated skirt women with gold bars on their shoes. I'd never wear anything out of fashion, but I also wouldn't wear anything in fashion that didn't suit me. Those jodhpurs, for instance – they're not at all flattering. Then it's not me wearing dropped waists or the Empire line. I hate them both. I hate polo necks too, but love stocks, though they never seem to be in fashion. Neat coats aren't my style at all. I'd never buy a shiny jumper, or wear skirts and cardigans. I *hate* tights and hardly ever wear them. Even in winter I go with bare legs and boots. Jewellery doesn't mean much to me. All my good stuff was stolen. I don't much like junk or baubles, but I love my diamanté cross and my Cartier watch.'

On the subject of Englishwomen in general, Susie is disparaging. 'They don't dress with the idea of *appealing* – they just put clothes on. I like to look at other attractive women, but I think they honestly don't *care* enough. They aren't oriented to males. It's a hangover from women's lib, perhaps. They just can't be bothered to snare men by looking yummy. I can't think of any Englishwoman who looks special, except the Princess of Wales, who obviously spends a lot of time and trouble on her clothes and looks wonderful and alluring, absolutely yummy. As my husband is so appreciative of how I look, it's him I dress for, especially when we go out. But as I say, I dress for *impact* . . . I love appreciation, I love to be noticed. Mentally clotheswise, I'd like to be the eternal mistress. . .'

Sue Lawley

BBC NEWSREADER

'On Saturdays there's the relief of not having to bother because seven or eight million people won't be seeing me. That's my escapism.'

Sue Lawley started as a 'Black Country bumpkin' – her own description – went to Bristol University then trained as a journalist. She began her career in television with the BBC in Plymouth, and moved to London and Nationwide in 1972. After four years as presenter and interviewer there, she worked for Tonight, and after another spell with Nationwide moved on to the Nine o'Clock News. For the past two years she has been with the Six o'Clock News four days a week. She writes many of her own links, headlines and end quotes, and does some interviewing. Appearing on live television still remains a nerve-racking business – the worst time is between half-past four and five in the evening – however used to it she is. Divorced, Sue lives in Fulham with her son and daughter.

Classic navy Cerruti suit which Sue says she wears more than anything else in her wardrobe. The white silk shirt with padded shoulders is also a classic, though 'slightly Dynasty*', and one of the most popular with viewers.*

'It's much more difficult to be well dressed than I ever thought,' says Sue, 'but clothes have always been important to me. My mother owned a ladies' and children's outfitters shop and I always remember her being well dressed. I would look into her wardrobe very appreciatively – Jaeger and Aquascutum things, plain soft green and navy suits and spotted shirts. Not many, but they worked and coordinated. My own taste is very classical, perhaps because of coming from that sort of background. My idea of heaven is a well-cut skirt with pockets and a lovely silk shirt.

'When I was young my dream was to have a classic cashmere coat with a tie belt. When I came to London, at twenty-five, it was the first thing I bought, for £100. It ought to have been a good investment, but it wasn't: there's no such thing as a *classic* garment because small things change – collars, the size of the cuffs, and so things inevitably date. When I was working for Nationwide, I began going to Elle and buying classic expensive silk shirts, and expensive trousers from Daniel Hechter. I thought they would last, but of course they didn't: they soon looked too small, too tight, too ungenerous. I don't read the fashion pages although I buy *Harpers & Queen*: fashion seems to affect me from a distance. Almost without realizing it each season my eye alters – with me, this happens months after the change has actually happened. I go into a shop and scoff at some new shape, then four months later find myself returning and asking for it. I was very up-to-date with the winged collar business, and keep on wearing winged collars because they suit me. If something currently fashionable doesn't suit me, I won't touch it with a barge pole.

'In fact, not many things suit me by any standards. I have a genuine problem: I'm big, and cheap clothes are cut skimpily. Good, large ones are always expensive. Working in television, I like my clothes to be dual-purpose: I feel guilty if I spend money on something I couldn't wear on TV. The most important thing to me is to be *comfortable*, and to be pleasing to the viewer's eye. I buy what I like and enjoy – the BBC only comment if they think something doesn't work. I wear a lot of white, though the lighting men complain like merry hell – I like white under the chin, and contrasting collars. My only rule is to try never to wear the same thing twice in one week. The letters I get from viewers are mostly complimentary, though some think I'm too severe. Why dress like a man? they say. Why not pretty yourself up a bit? I have quite a few letters from husbands who want to buy shirts like mine for their wives: there are two in particular that seem to be favourites with husbands.

'It's very difficult to find time to go shopping. But I do try approximately every spring and every September. I'm not as organized as I think I ought to be but I've got friends in various shops who ring me at the beginning of the season and say, hey, we've got something you might like, so I dash round knowing it won't be a waste of time. The Beauchamp Place Shop I often go to: I love their Cerruti suits and blouses. And Catherine Walker at the Chelsea Design Company, who makes to measure, has managed to get me into dresses. I have several pretty summer ones from her – her classic sailor suit in silk with a dropped waist, and two other cotton ones with dropped waists – a shape I'm very fond of. She has lovely evening dresses, too, the sort of thing I have to wear at a BAFTA-type do. I don't look good in many evening clothes: I hate fuss, frills and patterns. For medium-dressed evenings my theme always seems to be the same – frothy white blouse, black skirt (I'm into black skirts at the moment and have three) or trousers. I tend to be repetitive.

'There are various things I never wear: I hate the ethnic look, padded waistcoats, vast skirts, anoraks, the layered look, big jumpers with big

Sue has no great enthusiasm for evening dresses but bought this one by Catherine Walker at the Chelsea Design Centre 'simply because I liked it – it's just lovely to have'. Made of silk satin, she wore it presenting the BAFTA awards a few years ago.

20

patterns, and *Dynasty* things for daytime. I never wear synthetics although I was once conned by one that was successful – I like cotton, silk and pure wool. I'm absolutely not a frill person. I like necks either to be very high or very low – I can't wear *middling* anything. I have to have extremes. A collarless shirt, on me, is terrible: I like something sporty and easy, or high and neat, often with a big floppy bow. I'd never wear a T-shirt on television – for working I believe in being decent, crisp and clean. But at home I wear a lot of trousers and jeans. On Saturdays there's the relief of not having to bother because seven or eight million people won't be seeing me. That's my escapism.

'I can't wear colours,' Sue goes on, 'most things are neutral – beige, cream, white, black and navy. I do like some clear colours, though: a clear pink, for instance. In a muted pattern, or that murky green, I look washed out and terrible. Really, I'm a separates lady, and most of my blouses are cream, classic and silk. I do make mistakes. Sometimes a shirt that *I* love – like a favourite navy Jasper Conran one – doesn't work on television at all. But on the whole I keep things for years and have to force myself to have a clear-out every now and then.

'I don't care as much as I ought to about accessories. I resent paying a lot of money for a bag, so I own basics. I have a black Gucci evening bag and a St Laurent clutch bag. I'm a big clutch bag person. My great standby is a huge Gladstone bag which I bought from Harvey Nichols for £120 in 1970. It holds everything, and I've used it every working day and trip abroad ever since. I sometimes wear hats, but they're not serious hats. I bought a man's panama in 1984 which I wear with a cream Cerruti summer suit. I went to Ascot in that – I could never wear a hat with a veil or a flower. The only other hat I've ever enjoyed was from

Another classic wool suit by Cerruti, the coat cut like a man's dinner jacket. She had a bib and tucker put in so that she can wear it with nothing underneath. Both Sue's Cerruti are over three years old. Panama hat by Herbert Johnson.

David Shilling – it was vast, just a brim: for some reason I was photographed in it for the cover of the *Radio Times*.

'I do love shoes, but because my feet aren't seen it's very difficult to justify spending a lot on them. For a long time I used to get little else but Jourdan: they're too expensive now. I get classic courts from Russell & Bromley, and clip-on bows. I like high heels every day. My heyday was the time of ankle straps, T-bars and very high heels . . . I like lots of coloured tights, and tights with seams and big bows on the heel. Being such a classic dresser I suppose something has to escape, so I get tarty round the legs and feet . . . As for jewellery, I like classic imitation – I've some lovely imitation Cartier and Dior necklaces from Italy. I like big earrings, but I'm not really into jewellery: I'd never wear much on television, not wanting to distract the viewer's attention. I have a theory one shouldn't buy one's own jewellery or scent: my nice watch, rings and earrings were all bought for me.

'Who I dress *for* changes all the time,' says Sue. 'There are times when I definitely dress for male company. I'm aware there are some fashionable things that men definitely don't like. If I'm going to a business lunch with a man, say, I'd never wear flat shoes. In the evening I dress for the male eye, but I know that on certain days things will attract females rather than males, and I like it when girlfriends say "Oh, that's terrific!" I admit I enjoy *owning* my clothes. It gives me a sense of well-being knowing there are beautiful objects in my cupboard.'

Sue's talent seems to be an eye for the delightful classic rather than the dull classic. For the purpose of her limelight job she has to have an unusually large number of clothes, but the outside observer would find it difficult to spot a mistake in her crowded wardrobes: they are full of things 'pleasing to the eye'. As one who surely deserves to win any poll for being the most consistently elegant woman on television, her views on the English way of dressing are not encouraging. 'I'm ashamed of us,' she says. 'We just don't care enough. We just don't take enough trouble. The Princess of Wales is an amazing exception: one of the few people whose clothes I admire. But my long-time heroine is Lauren Bacall – I interviewed her once. I shall never forget that long straightness, that faint rustle of silk lining. I think Charlotte Rampling is wonderfully dressed, too: our generation's version of Bacall. But it's as a Lauren Bacall dresser I'd like to be remembered – the rustle of silk as I walk . . .'

Laura Gregory

FILM PRODUCER

'I like to be a bit outrageous. At the Cannes Advertising Film Festival last year I was jokingly awarded the prize for the best-dressed advertiser – I wore a naughty dress of leather and chains.'

Laura found this silk dress 'so vibrant' she could not resist it, for all its expense. From Trashy Lingerie in Los Angeles: matching peacock shoes. Laura would wear this all day at the office if she was taking clients out in the evening and had no time to change.

At twenty-eight, Laura Gregory's career has been genuinely dazzling, high powered, the stuff of modern success achieved through enormous energy and total dedication to her job. Daughter of Antoinette Gregory, the theatrical costumier, Laura began as a market researcher. She then joined a Bond Street jeweller as a rep, and went on to be a receptionist at a production company. She worked her way up through the company and, in love with advertising and the making of commercials, started her own very

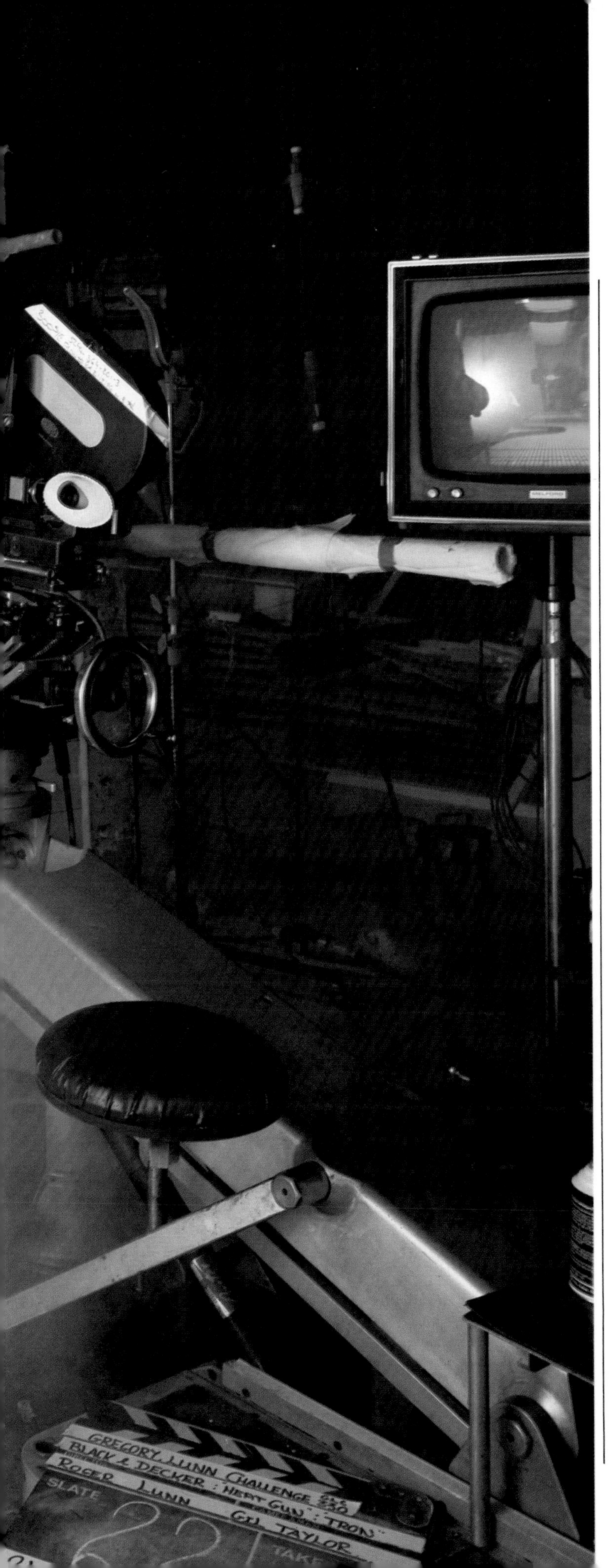

LAURA GREGORY

successful company, Gregory Lunn Challenge, seven years ago. With three producers and four directors, she makes films for advertising companies. 'Basically a salesman', is how she describes herself. Much of the job involves jet-setting – thirty countries visited in 1985. She lives in London, loves the ballet and opera, starts the day with a workout at The Fitness Centre every morning at seven o'clock, and, seemingly indefatigable, survives on very little sleep. In June 1986 she married the photographer Ken Griffiths.

'Advertising is a celebrity business, it's all about being noticed, about selling,' says Laura. 'Most of the women in it are between twenty and thirty and take a lot of time and trouble over their clothes. Working so much abroad, their outlook to dressing is probably more European than English. We're lucky in that in our industry we can wear what we please, whereas a lot of women in offices have to dress in a certain style: they don't have the choice I have. The Englishwoman in general seems to me pretty drab. She likes drab colours. In France and Italy women have a knack for accessories – some little thing that makes plain clothes special. There's not much of that here.

'I've always been a *theatrical* dresser. Even at school I was a bit of rebel. I wore the regulation skirt down to my ankles, with Granny shoes instead of the official ones. Up till the age of sixteen my mother made everything for me. Even now I sometimes like to design my own things and she will cut

Dark brown leather jodhpurs and cream silk shirt from Hilary Moore in Covent Garden, both 'fairly new acquisitions' and typical of the comfortable, practical sort of things Laura wears while on a shoot for an advertising film.

me a pattern. Today, every commercial we do brings a query as to how people should be dressed: stuff is brought in all the time which I look over for myself as well as for the film, and so I'm always aware of what's around. Models come in – I see what they're wearing and ask where something comes from if I like it. I do keep in touch with fashion, but I wear what *I* enjoy wearing. I probably buy something every week, thought I don't have much time for shopping. I'll see something in a magazine and ring up the store. I *never* try clothes on – if it's the right size, I'll buy it; if it turns out to be a mistake, I'll give it away. Whenever I go abroad I usually manage to fit in a couple of hours' shopping. I buy silks in the Far East, shoes in Italy. New York is the most fun place to look around. I do spend a lot on clothes. But they're part of my personality, part of the job. At awards ceremonies, I'm always asked what I'm going to be wearing.

'I like to be a bit outrageous,' admits Laura, who has achieved fame for her clothes in the advertising world. 'At Cannes last year I was jokingly awarded the prize for the best-dressed advertiser – I wore a naughty dress of leather and chains. It came from Trashy Lingerie in Los Angeles – a lot of film stars go there for their lingerie and some evening dresses. Joan Collins and Cher are customers. They keep your size on file, so when you need more pants they send them over. They're more expensive than Marks & Spencer, of course, but more fun. I'm mad about shoes, I've got at least two hundred pairs. I spend ages hunting down exactly the right colour. I like very plain court shoes, and boots. I've found the best ones here come from Geiger, Bertie, Midas, Hobbs and Ad Hoc. Tights don't exist in my vocabulary. I always wear stockings, in every colour, silk when I can find them. I search for them all over the world.'

Laura claims to buy 'rather classical' clothes, although her idea of classical might not match the average concept. 'I was brought up to believe that it's better to buy one skirt for a hundred pounds than three for thirty,' she says. 'I admire the chic, tailored look – the Armani look, but think you have to be very thin. Not for me. My mother has collected antique dresses for years and I have some lovely ones – 1930s dresses cut on the bias, satin and laces: but I don't like 1960s things. My wedding dress was a copy of a 1957 Dior dress, in blue silk. I'm a great recycler, I'll wear something for three or four years, put it away for six months and then bring it out again. I'm also an Aquarian, and Aquarians are bizarre dressers – I could even go for a twinset and pearls one day, thought I haven't yet. I find myself going a bit cowboy in America – coloured stetsons, pointed lizard skin boots and so on.

'I've always been a sports freak and used to do a lot of swimming, so have big shoulders. I like to make a feature of them. Parachute in New York is one of my favourite designers – great fun for working clothes – huge shoulders on jackets, the same for men as women. In Milan Jean-Paul Gaultier is my favourite designer – wraparound dresses with knitted bosoms. He's got real wit. In the office I mostly wear trousers, or a leather boilersuit, or Maxwell Parrish jodhpurs: I like to feel relaxed and unencumbered. I *love* leather and suede – if you saddle soap leather it gets better every year. When I'm filming on location and it's very cold I wear a wonderful Alli Cappelino over-sized sheepskin coat. I love hats and buy them anywhere. There's a wonderful shop in Monmouth Street where Colin Swift, who's an artist, makes me pillbox hats. A particular favourite is red velvet with ostrich plumes. Paul Smith is a designer I'm very fond of – I have a favourite powder-blue bellboy jacket from him: and I love Japanese designers, Kansai and Issy Miyake. I've a stunning outfit of scarlet baggy trousers, waiter's jacket and bandleader's hat from Kansai. Travelling so much has taught me only to take light things that pack well – I choose things of just two colours, it makes life much easier. An especially good dress for travelling is a black Lycra tube, nine inches wide before it stretches to go over you – *very* clingy. At home I wear mostly interesting track suits. One of my favourites is from Miyake – £300 it cost, but it's amazing, perfect to travel in and will go on for years.

'I like all colours but brown – royal blue is my favourite – and all natural fabrics, especially cotton and silk. I'd never wear synthetics except viscose, which really can feel like silk. I think there's an art in knowing what suits you: what shape – I love huge shoulders and small waists, jumpsuits, ski pants, big roomy shirts. I also love ball dresses, though I don't own one: if I have to go to an advertising ball I hire one for the evening from Bermans, any period. There are lots of things I hate and would never wear – modern glitter, modern sequins, and lurex; knitted dresses and jumpers, kilts and tweeds. I don't make many mistakes in what I buy, but I sometimes choose the wrong things in the morning and it makes me really bad-tempered. If my clothes don't feel right it spoils my whole day.

'The trouble is that I often have to dress in the morning in something that is not only going to carry me through the day, but in which I have to take clients to the opera or ballet in the evening. It might be, say, something like my gold brocade jodhpurs, with waistcoat and jacket and oversized silk tailshirt, and matching fez hat. I like wearing a hat at night – occasionally I wear a top hat with my dinner-jacket. I don't much aspire to jewellery. I love my waterproof watch but otherwise all I wear are two diamond and sapphire studs in one ear, four in the other.

'Ken likes my clothes,' says Laura, who describes herself as an outrageous dresser, 'but I dress for *me*. I love people to notice the effort I've put into my appearance, and I certainly notice what other women do. The person whose dressing I most admire is Annie Lennox of the Eurythmics. Her clothes are very Marilyn Monroe, very feminine. Clothes will always be important to me, though I don't talk about them. I just do what pleases me – they're an extension of my personality. For advertising parties I really enjoy dressing up. I'll go much more outrageous – to please myself – than I would for taking clients to the opera – I'll wear wild wigs, rubber dresses, leather and chains. One of the most important things about dressing, I believe, is to keep a sense of humour. . .'

Previous page: 'Fantastic brocade and leather boots to go with a simple silk blazer, pleated skirt and silk vest from Hiroko Kishino, a Japanese designer in Milan who sadly has no shop in London. Pillbox hat from Colin Swift. Another combination Laura would wear all day, adding the hat in the evening.

Maggie Thompson

FABRIC REPAIRER

'I'm on the look-out wherever I go, mostly in charity shops. Often I find something revolting but buy it for the buttons or belt or the fabric ... texture and construction of old fabrics are what I really love.'

Maggie Thompson comes from a long line of craftsmen of the old school. Her grandfather was a wheelwright and her father a carpenter. She herself went into book-keeping but discovered she was good at mending things 'so that they looked as if they haven't suffered too badly'. The restoration of some old Chinese wallpaper was her biggest triumph, and she miraculously mends antique carpets and curtains that most people would consider beyong repair. Her husband is a trained cabinet-maker and restorer. They have a workshop in Marlborough and live nearby with their two children.

'I don't like shops,' says Maggie, 'the loud music and general presentation. They're never quiet enough for you just to be able to go in and *imagine*: lights and security and everything add to the panic – I don't like all that. I get very tired and can't find anything, unlike my husband, Barrie, who's an expert in modern shops. He can track down anything he wants straight away. I think you should be able to snatch something off a shelf and create a feeling, but you can't do that now. I do go on a reconnaissance mission from time to time, more for the children than me. *Sometimes* the buyer gets a colour right, but not often. I don't spend much time thinking about *fashion*: what I like are clothes from the past. The texture and construction of old fabrics are what I really love, things from the twenties, thirties and forties, which you're likely to find in charity shops. There were wonderful fabrics in those days. It's *fabrics* I hone in on, they're what's *most* important to me.

'I think I probably spend very little on clothes; we were so hard up, starting the business, I got in the habit of spending very little, and can't break it. I'm on the look-out all the time, mostly in charity shops. I go into the Thrifty Orange in Marlborough – proceeds go to the Alliance – several times a week. Often I find something revolting but buy it just for the buttons, or belt, or fabric that I can make into something else. I've been very lucky in the past, though there seems to be less good stuff nowadays. Once from Oxfam I found a nineteen-twenties chrome and plastic belt – the engineering was incredible. Folded up, I though it was a box. A lot of the my clothes come from a friend's rag-bag – she throws out things we can make into rags for the workshop. I found a beautiful black crêpe dress, about 1960, I would think. Sometimes I wear it with a nineteen-thirties olive-green suede stitched belt, or a thirties rag-bag jacket of silk velvet in beigey-greens. But the most wonderful find of all from the rag-bag was a black velvet coat lined with rust silk – about 1910. I love things like that you can dress up in and be silly: so far I've only worn it once.

'For everyday I wear trousers and jumpers and shirts – I feel most comfortable like that. I get my trousers from a shop in Marlborough. Once – it was like Christmas Day – they had *mountains* of trousers – three pairs for five pounds or six pairs for ten pounds. No hangers, no rules – I dived in and came out with six pairs in colours no one else wanted – rich magentas, blues, peacocks: they were quite the wrong shape for *fashion*, which I don't follow. I've got a favourite dark blue man's shirt, very fine cotton, from a Sue Ryder shop, and I wear a Dr Deimal thermal shirt with a collar that I found at a jumble sale. I dyed it pink and wear it under beautiful thin cotton shirts – it acts as a vest. I also

Maggie and her husband share a pole rather than a wardrobe, and Maggie's share is the smaller. But then she has few things that do endless service. The black crêpe dinner dress came from a rag bag, the silk jacket from a jumble sale twenty years ago. 'They're very comfortable, things I feel easy in if we're going to some special dinner or theatre,' she says.

love and often wear a nineteen-thirty-nine blue jersey, which I keep mending and mending.

'Colours I like used to be browns and greens, but that's passed by. Now it's blues and amethysts – off colours, all the in-betweens. I'm not mad about yellow or coral. I've a lovely, huge, linen and wool cardigan – it's made up of lots of flecks but the overall effect is lavender. I got it from a friend so I could try it on in the workshop. It's functional – I think clothes have to be. They have to work. For me, if clothes don't work, aren't comfortable, then *I* can't function . . . I don't like clothes that are just for show, and it disturbs me if people look uncomfortable in what they're wearing – their shape gets affected. I have not to think about what I'm wearing once I put it on, so that I can get on with work.'

Surprisingly, for one who finds fabrics so important, Maggie is fond of some synthetics. 'I love viscose,' she says, 'if it's good quality, very soft. But it has to be the right design and colour. It creases like the devil and needs a lot of ironing, but I like wearing it. I like rayon, too, and then natural things – cotton and wool, sometimes silk – though not dressing-gown silk. I'm not mad about satin and velvet unless they're old. As for shapes – no real rules, depends on the mood. I'm not keen on frills and flounces. I like something that can be drawn in at the waist, but I also adore drop-waist garments, things that look as if they've been *poured* on, that clinging look that just comes out at the bottom. I love T-shirts, loose fitting, and old Aertex sports shirts are great favourites. I don't have anything conventionally suitable for, say, a dinner-party, but I'll usually manage to find something in the dressing-up box that will do. Over the years I've collected *drawers* full of bits and pieces. I often have to be sewn or pinned into something, but I'm not inhi-

Elizabethan crewel work is the sort of fabric Maggie Thompson finds herself repairing in her workroom in Marlborough, and variations on corduroy trousers-with-shirts-and jerseys are her daily uniform. This amethyst flecked cardigan is a particular favourite – found by a friend in Monsoon, and passed on.

bited in that sense, I have no qualms. In fact, people comment on my clothes, compliment me – I can't believe it! But it all seems so easy: it's second nature to me to be on the look-out for this and that.' (Robert Kime, the antique dealer for whom Maggie works, says of her: 'Maggie is a chameleon. You could put her anywhere in the world and she would fit in. Instinctively, she would choose the right clothes.')

'I haven't much jewellery,' says Maggie, 'but wear mostly a selection of things that fit in with the fancy-dress wardrobe – beads the children have made from rolled-up copies of the *Sunday Times*, gipsyish things, a pretend-silver snake that started off as a dog collar. I throw on whatever jewellery goes with my mood, usually old stuff.' Her collection includes an exquisite paste arch, some two inches high, from which dangles a paste frond. Probably circa 1930, Maggie wears it as a brooch. 'Shoes?' She smiles. 'I have one pair of 1964 leather court shoes: they get dyed whatever colour is needed. At the moment, they're mulberry. For every day, I wear fifty pence sneakers from somewhere.'

Maggie's view of the Englishwoman's dressing is that it's not particularly dowdy, but the majority of clothes available are generally pretty boring. 'But when I look at other women's clothes I only notice things I like,' she says. 'I'm fascinated by people who've made up their minds and become wonderfully stuck in a time warp – someone, say, who wears long seventies Indian dresses and is as happy as Larry. What I love is Zandra Rhodes's approach. For her, nothing is impossible. She's the only designer I can think of who comes close to capturing all moods and putting them into one garment. One of the best-dressed women I know is Helen Nicholl (author of the *Meg and Mog* children's books). Helen always looks marvellous: yes, she and Zandra Rhodes I admire.

'I find this difficult to admit,' Maggie says, 'but I do basically dress for myself. My husband only says something if I don't look good, and I can be affected by that. I might not change what he doesn't like, and still go out in it, but I remain affected. Sometimes, surprisingly, he doesn't say anything at the time, but comments *after* the event – something like: "Why don't you wear that thing you looked nice in for such-and-such an event?" But he doesn't often say much: that's Barrie. What sort of dresser am I?' Maggie reflects for a long time. '*Inconsistent*,' she decides eventually. 'I think I change all the time to fit the mood. I suppose what I should really have been is a wardrobe mistress, because what I *most* love is to dress up other people.'

JANE STEVENS

MARKETING DIRECTOR OF JACQMAR

'To be well dressed I think there has *to a touch of drama – though of course there's a thin line between that and vulgarity, which you mustn't overstep.'*

For lady-in-waiting duties or important daytime occasions: plain wool dress under a check coat with a velvet collar, by Hardy Amies, and incorporating the 'touch of drama' Jane thinks is essential to elegant dressing.

Jane Stevens, it could be said, has perfected the art of organization. With extraordinary energy she runs a beautiful house near Abingdon, and a London flat. Her whole-time job is Marketing Director of Jacqmar, for whom she is gathering together a collection of accessories to be sold nationwide, which takes her all over the world. She is also a part-time lady-in-waiting to Princess Margaret. She has three grown-up children, a daughter of fourteen and, amazingly, three grandchildren. At weekends she provides some of the most comfortable and merry times imaginable for her friends: the epitome of English country-house life that still exists in a few places. But even at weekends, she herself doesn't stop, getting up as usual at dawn, however late she has gone to bed. Her claim that she has little time to think about clothes seems reasonable: the puzzle is how she manages to be so consistently elegant and maintain what is surely one of the most impressive wardrobes of any Englishwoman.

JANE STEVENS

'I've always been an impulsive buyer,' says Jane, 'I never go out and get something for special occasions – I buy because I've seen something lovely and then think the occasion might arise. I'm very economical, though that's not what others say,' she adds with a smile, and indeed a glance at her myriad clothes makes one think it must be a very individual kind of economy that she practises. 'Because I have to think about every penny, I often go to sales – Belinda Bellville, for instance. I went to Hardy Amies quite by chance and find his ready-to-wear collection quite remarkable, and very reasonable. His things are in beautiful materials and well made – trouser suits, skirts, a lot of mix and match, though I hate that description. I've got some lovely silk trouser suits for the evening from him, and an ivory silk skirt and shirt, and a wool and leather jacket to go with jeans that will never date. I love beautiful materials and will never, ever, wear man-made fibres. I've been going to Gina Fratini for many years – a new favourite from her is a long purple velvet dress with a huge Puritan lace collar. I've also been buying things from Jean Muir ever since she began. When she had an exhibition that toured the country some years ago, I lent her a lot of her clothes that I'd had for years. I think Jean is the most classic, long-lasting designer we have.'

Among her myriad activities Jane sells a selection of T-shirts, bags and exotic capes from her London flat. This T-shirt is one of many she has especially painted.

The Englishwoman's Wardrobe

For ages Jane has been reflecting on an idea which she hopes might one day materialize. 'I really wish there was one huge trade fair a year, rather like the Decorex interior decorating one, in which everyone showed some of their things in small booths. What is produced in this country is often wonderful, but people don't know where to find it. A huge, annual exhibition with everyone displaying their wares under one roof would be marvellous. As it is, I do go to some trade fairs and like discovering new designers at them. For instance, I've found some lovely things from Wim Hemmick, a Scandinavian, and Jacques Azagury. I also discovered Lindka Cierach long before Sarah Ferguson's wedding dress made her so famous. I've been going to her for twenty years. From her, Jane has a very romantic blue and white satin evening dress with a vast collar.

'I do definitely like a bit of drama in clothes,' she says. 'Most Englishwomen go for the undramatic: I tend to be the opposite. To be well dressed there *has* to be a touch of drama – though of course there's a thin line between that and vulgarity which you mustn't overstep. I've always loved bright colours, which again Englishwomen don't, on the whole. I love red, electric blue, black, yellow, apricot – though *never* green. I'm a classical dresser, I say quickly, though also, sometimes, a romantic one. I wear a lot of smart trouser suits, ordinary suits, skirts and shirts, but hardly ever day dresses. I find I buy plainer and plainer clothes with wonderful accessories, beaded bags and beautiful belts. I find myself wearing the same shape a lot – perhaps a sign of getting old.

'I love beautiful long dresses and keep some of them for twenty years or more,' Jane goes on. They keep their pristine condition because they are exceptionally well cared for. Jane's various wardrobes are arranged with military precision into kinds and colours of clothes. As all other areas of her life, her cupboards are highly organized, ultra-tidy: and yet that state seems to come naturally to her. It's achieved with no fuss. 'I don't *just* go to expensive places to find things,' she says, further to prove her sense of economy. 'I once bought a black taffeta evening dress with velvet spots from a shop called Fred's in Smith Street – twenty-five pounds and one of the prettiest things I've ever seen. I wore it at Covent Garden and everybody thought it was from Paris. I go to Marks & Spencer for underclothes and found some wonderful leather and suede trousers there. I was terribly excited when I found The Sock Shop at Paddington Station: stunning socks in pretend tartan that you used only to be able to get in Geneva. I bought eight different colours. I *always* keep an eye open.'

In any spare time Jane visits country antique shops where this beady eye alights upon things to add to her collection of antique jewellery: Victorian and Edwardian paste buckles for belts and shoes, and butterfly brooches. Her jewellery is mostly semi-precious – lapis, onyx, jasper and so on, often in the form of very long necklaces. The success she has in finding antique jewellery, though, is not matched in the area of shoes. '*Where* can you get pretty evening shoes?' is her constant cry. The romantic spirit in her yearns for the kind of old-fashioned evening ones it seems impossible to buy: her solution is to have them made in velvets and brocades and decorated with her paste buckles. For day, she likes Italian shoes: Pied à Terre in Sloane Street is a favourite shop. Tights she wears in all colours, either contrasting or matching her clothes. 'As for hats, I wear them when I have to, and get most of them from Graham Smith. I love romantic hats, and have quite a collection, lovely straw ones with bows and veils, many of which I've had for years.

'I never think I have enough clothes because things change so all the time: but everything comes back eventually, so I'm a great storer, and will happily take something out that's been put away for twenty years. I'm very aware of current fashion: I care about it and take it all in. Having a young married daughter and a teenage one – both very fashion-conscious – I'm aware of what the young are into and what's going on. My fourteen-year-old, Melinda, is a very meticulous dresser, she has an extraordinary eye. "Let *me* choose what you're going to wear," she says, and she's always right. Everything of hers is loose and large, pointed toes with flat heels – I can't buy for her. Although I'm always very in touch with what's in fashion, I don't *talk* about it. Subconsciously I seem to have a nose for what's coming in, colours and so on, and I find I'm always ahead of everyone through some sort of instinct. Because of the life I lead, clothes are obviously very important and, luckily, finding them comes naturally to me. I very seldom make an expensive mistake, though sometimes a cheap one. I've never, ever, bought something expensive, worn it once and regretted it. I know myself: I'm never untidy, for example. What I loathe is the punk look. I think it's hideous, dirty, seedy. I think one should make the *best* of whatever one is. Luckily, I don't have to worry about putting on weight – my figure never changes which makes it a lot easier. But,' she adds, 'I do have elastic waists on many skirts which is a marvellous solution to slight fluctuation. . .'

Jane admires very few Englishwomen's clothes. 'As classic dressers I admire Princess Alexandra, and my sister-in-law Prue Penn. Also Diana Heinemann, who has great bezazz. The person I *most* admire is Anouska Weinberg: she always makes an impact with something that's completely her own style – dotty, amusing, slightly eccentric and sexy. But then she's half German, half Russian. . . On the whole I think Englishwomen's dressing is *terrible*. What they simply don't know is how to put it all together. They don't know how to walk, or enter a room or do their hair, even, apart from actual clothes. And *hair*, I think, is one of the most important things – again, something Englishwomen don't bother with. If your hair always looks good you can get away with slightly less perfect clothes. If you make the effort to have it done to go under a special hat for, say, a wedding, it will make all the difference. Then Englishwomen are always looking for something cheap. They lack courage: they don't *care*. Look at a French shopgirl compared with an English shopgirl. . . . My belief is you should forget what husbands think, and *dare*. I don't think there's anything typically *me* about my clothes: they're very disparate. I certainly don't dress for men, or for women – I dress the way I do because *I* like it. To be honest, I do like appreciation for the way I look' – and naturally she gets a great deal – 'because I've made an effort. The day you give up trying, then forget it. But if you go on caring, and have a sense of humour, you can get through anything. I firmly believe that applies to clothes as well as to life in general.'

Previous page: After a twenty year rest this silk crêpe evening dress with matching ostrich feathers has had a recent renaissance. It was designed by Pierre Celeyron for a shop called Oh! which Jane and two partners owned in the late sixties. She first wore it to a ball at the British Embassy in Paris given by Lord and Lady Soames.

Elaine Bennett

THEATRICAL DRESSER

'A tourist asked me why I dressed like this. I said I was carrying on the old English tradition of eccentrics. I'm lucky enough to have a big personality which I can exaggerate through clothes.'

Elaine Bennett is totally loyal to designer Jane Kahn, from whom she buys all her clothes. This blue lycra dress – a rare escape from black – she wears with the matching headband 'at any time of night or day'.

Elaine Bennett, aged nineteen, lives with her parents in a flat in the Fairway Old People's Home, Watford, where her mother is second officer in charge. Her father is a cyclist and she has an elder sister. Elaine left school at sixteen and, except for a few 'horrible little jobs', has been mostly unemployed. She joined the Watford Youth Theatre last year and had a part in a production at the Palace Theatre. But she does not want to be an actress. Her break came early this year when she was offered a job as dresser to Christopher Timothy when the Palace put on *The Real Thing* – her first job for a year and a half. This was followed by dressing Dorothy Tutin in Maria Aitken's production of *Are You Sitting Comfortably?*, again in Watford. Elaine describes herself as a 'very laid-back person with no big goals in life', but being a dresser in the theatre is the job she would most like to do.

'I began dressing as I do now in 1982,' says Elaine. 'I was a regular schoolgirl then, very drab, very boring. I can remember admiring the punks in the King's Road, and walking round the Gear Market. It was there I met Jane Kahn. Her corner in the Market was so colourful, so over to the top, so extravagant, I thought: "That's me." Jane is a wonderful designer. She herself dresses so shockingly it's pleasing to the eye. In her clothes I found an outlet where I could explode my feelings. So I began my own style: a cross between Barbara Cartland and a punk, some people say. But I'm not a punk, or a new romantic. The new romantics started dressing up in reaction to the punks dressing down,' she explains, 'but I haven't a clue what's going on now. People try to put me in a category, but I'm on my own. Last summer a tourist asked me why I dressed like this. I said I was carrying on the old English tradition of eccentrics. But is eccentric the right word for my dressing? Perhaps it's more *flamboyant*. I'm lucky enough to have a big personality which I can exaggerate through my clothes. I've no friends in Watford who dress like me, and I don't mind what others wear. It's just that my personal preference is flamboyance.'

Since her discovery of Jane Kahn, Elaine does not bother to shop in many other places. 'I look around Help the Aged and charity shops,' she says, 'and I bought all my fox furs very cheaply from them – one for five pounds, now converted into a hat. If I had lots of money I'd go to Bruce Oldfield: I like some of his stuff.' Elaine's normal gear is a dress that she uses as a base on which to display her myriad jewels, belts, chains, collars and so on. But there are some accessories that hold no importance for her. 'Shoes I don't care about as I don't think people are going to look at my feet,' she says. 'In the High Street chains they're the same in every shop, so I'll pop down to London and walk about Kensington Market till I see something I like. I'll bung on old tights. I don't care if they've got holes. At home I'll walk round all day in a nightie and dressing-gown and no make-up: the gear is only for going out. I have been known to throw on the make-up just to go down to the shop for a loaf of bread.' While her hair goes from silver-grey to white, Elaine's make-up is elaborate, exotic, a daily work of art. 'People say it must take you hours, the sequins being so symmetrical and that,' she says. 'But in fact I just sit there and boom, boom, boom the make-up comes – takes about a quarter of an hour, even with sticking on the sequins and jewels, which was another of Jane's ideas. Jane,' Elaine goes on, 'always gives me credit for the way I look. She helped me to grow up just by talking to me. She's a wonderful woman. She's like even wilder than me, with this wonderful figure. I often say to her I wish I was thin, but she says it doesn't matter. You often see Samantha Fox and Page Three girls wearing Jane's things when they're dressed up.'

Elaine is philosophical about the expense of Jane Kahn's clothes. A favourite detachable black collar, black padded stuff embroidered with rhinestones and trimmed with ostrich feathers which cost sixty pounds was, she found, irresistible. 'I have to save very, very hard from whatever I manage to earn, and from presents from my parents,' she says, 'but it's worth it. I only go on a shopping spree about four times a year. Soon as I've bought something I put it in a big bin: I might not get it out for ages. I made a deal with myself that I'll never throw anything away. When my kids say to me, Mum, you can't ever have been young, I want to say to them, Look!'

Unlike most people of her age, Elaine has never worn jeans in her life. 'They wouldn't suit me, being big,' she says, 'and I would never wear an overcoat or a mac or any kind of uniform. My great passion is *feathers* – huge feather hats – they don't half keep your head warm – feather trims, everything. I saw

Molly Parkin on television covered in feathers, even feather bows: she looked great. Then I love gold – anything that sparkles is real to me. There's a shop in Luton sells chains for toys, very cheap, so I bought tons of them. Black is my favourite colour. I either wear black or white. I only have one coloured dress – a blue. I like any material, but specially satins and silks: wish I could afford them. But I've got a wonderful stretchy black dress made of Lycra. I *love* big shoulder pads, and ball gowns. Jane made two for me, special order, but I very rarely get them out of the cupboard. Once I dressed up in my favourite one – black, with purple glittering stuff on the skirts with black net over *that* – and the old ladies up at the Home said, "My God, what's that?" But I think it cheered them up. They called me Scarlett O'Hara. Well, I want my clothes to give people pleasure, to make them happy. My whole life is based on fun. If I can laugh at myself, I can laugh at others. But I'd never say anything to hurt anyone, though others hurt me.

'I'm a very shy person,' Elaine goes on, 'so to wear this gear is a front to hide behind. I can be as loud as I like. I've always been big since birth, and I've always been put down. When I started to dress weird, people said, like, there goes the fatty in the strange clothes. But at least they're shouting at the charade, now. Without it, they'd be shouting at me, the fatty. I'm happier now, dressed like this.'

Elaine's parents have different opinions about her clothes. 'Dad hasn't really come to terms with them,' she says, 'but I think Mum likes them. It dawned on her that really I was doing the same as

Elaine can go wild with feathers – they're her 'great passion'. An ostrich feather collar eccentrically enlivens a simple batwing dress: the huge 'head-warming' hat is made from a less expensive bird, the cockerel.

KAHNIVEROUS
DESIGNER
JANE

AIR
CANADA
CARGO
LONDON MIDLAND AND STEAM
Scotties

she was doing back in the fifties with her big circle skirts and all those net petticoats. As for my sister – well, put it like this: if I bump into her in the High Street, she won't speak to me. She's ashamed of me in public. She's tall and slim and classical – she can get away with classical things. But no one could influence me about my clothes. I don't have a boyfriend, and if I did I wouldn't pay any attention if he criticized me: I definitely dress for myself. Perhaps this kind of dressing will end one day, I don't know. Maybe I will go on to become an old eccentric – so long as I can keep on getting the feathers. I'm not really aware of causing attention any more. To begin with, friends used to be embarrassed by me, but they're used to it now. People I don't know in the street shout, "Look at that old bag, that fat cow", but I just ignore it. And I don't *ever* get fed up with dressing up. No matter how much I get put down, there are always one or two people who will come up and say I look wonderful. I like that because it shows *someone's* noticed all the trouble I've gone to, they've taken it in and like it. But the theatre is probably the only place I'll be able to get a job, looking like I do. People in the theatre are very nice: they get to *know me* behind the clothes. And it's the personality behind the clothes that's important.'

The events of current fashion hold little interest for Elaine. 'I glance at magazines like *Just Seventeen*, *Etcetera* and *The Face*,' she says, 'but I don't take any notice of them unless Jane is in them. She's the only person whose clothes I admire – she's my guru. In general, I think clothes for young people are incredibly boring, especially for, like, the kids who have just come out of school and pop into Miss Selfridge, and all come out in uniform. They have no individuality, they're all identical – that's frightening. I admit I think quite a lot about clothes, but more about make-up. The way I dress and look now wasn't the result of any great master plan, it just happened. I can't imagine any alternative way of dressing: I have my principles. When I go out I *like* to make a spectacle of myself, even though people often think I'm an alien from outer space.'

Previous page: Best ball dress, made for her especially by Jane Kahn, earned her the nickname Scarlett O'Hara. The sequinned make-up began early this year and will not change 'till inspiration runs out'.